BIOTECHNOLOGY IN AGRICULTURE, INDUSTRY AND MEDICINE

BIOMETRICS: THEORY, APPLICATIONS, AND ISSUES

BIOTECHNOLOGY IN AGRICULTURE, INDUSTRY AND MEDICINE

Additional books in this series can be found on Nova's website under the Series tab.

Additional E-books in this series can be found on Nova's website under the E-book tab.

BIOTECHNOLOGY IN AGRICULTURE, INDUSTRY AND MEDICINE

BIOMETRICS: THEORY, APPLICATIONS, AND ISSUES

ELLEN R. NICHOLS
EDITOR

Nova Science Publishers, Inc.
New York

For permission to use material from this book please contact us:
Telephone 631-231-7269; Fax 631-231-8175
Web Site: http://www.novapublishers.com

NOTICE TO THE READER

LIBRARY OF CONGRESS CATALOGING-IN-PUBLICATION DATA

Biometrics : theory, applications, and issues / editor, Ellen R. Nichols.
p. ; cm.
Includes bibliographical references and index.
ISBN 978-1-61728-765-7 (hardcover)
1. Biometry. 2. Biometric identification. I. Nichols, Ellen R.
[DNLM: 1. Biometry. 2. Biometric Identification. WA 950]
QH323.5.B545 2010
570.1'5195--dc22

2010026083

Published by Nova Science Publishers, Inc. † New York

CONTENTS

PREFACE

In recent years, numerous research papers have considered the problem of privacy protection in biometric authentication systems. When someone wants to authenticate himself, the fresh biometric data he presents has to match with the reference related to the one he claims to be. Many proposals have been made to ensure the confidentiality of the biometric data involved during this verification. Also discussed in this compilation, is the problem of finding a face recognition system that works well both under variable illumination conditions and under strictly controlled acquisition conditions. Information fusion techniques are also explored and have been widely developed in the community of voice biometrics along with several uni-modal speaker recognition algorithms.

Experimental clinical trial settings are now often extended to community entities beyond academic research centers. In such settings, a cluster randomized clinical trial (cluster-RCT) design can be useful to rigorously test the effectiveness of a new intervention. Investigators are most commonly interested in assessing the following three types of intervention effects: overall intervention effect, change in intervention effect over time, and local intervention effect at the end of the study. At the design stage of the cluster-RCT, it is essential to estimate a sample size sufficient for adequate statistical power to evaluate the different intervention effects. However, the sample size estimation must account for the multilevel data structure that is necessitated by the nature of the cluster-RCT design. In Chapter 1, the authors consider a three-level data structure and summarize sample size approaches for testing intervention effects within a unified framework of mixed-effects linear models which offer flexibility in the analysis of multilevel data and hypotheses testing in a cluster-RCT. The sample size methods are presented in closed form and have been validated by simulation studies. Important features of sample size determination for each primary hypothesis are also discussed.

One important application of gene expression profiles data is tumor classification. Because of its characteristics of high dimensionality and small sample size problem, and a great number of redundant genes not related to tumor phenotypes, various dimensional reduction methods are currently used for gene expression profiles preprocessing. Manifold learning is a recently developed technique for dimensional reduction, which are likely to be more suitable for gene expression profiles analysis. Chapter 2 will focus on using manifold learning methods to nonlinearly map the gene expression data to low dimensions and reveal the intrinsic distribution of the original data, so as to classify the tumor genes more accurately. Based on Locally Linear Embedding (LLE) and modified maximizing margin criterion (MMMC), a modified locally linear discriminant embedding (MLLDE) is proposed for tumor classification. In the proposed algorithm, the authors design a novel geodesic distance measure, and construct a vector translation and distance rescaling model to enhance the recognition ability of the original LLE from two aspects. One is the property that the embedding cost function is invariant to translation and rescaling, the other is that the transformation to maximize MMMC is introduced. To validate the efficiency, the proposed algorithm is applied to classifying seven public gene expression datasets. The experimental results show that MLLDE can obtain higher classification accuracies than some other methods.

Voice biometrics, also called speaker recognition, is the process of determining who spoke in a recorded utterance. This technique is widely used in many areas e.g., access management, access control, and forensic detection. On the constraint of the sole feature as input pattern, either low level acoustic feature e.g., Mel Frequency Cepstral Coefficients, Linear Predictive Coefficients or high level feature, e.g., phonetic, voice biometrics have been researched over several decades in the community of speech recognition including many sophisticated approaches, e.g., Gaussian Mixture Model, Hidden Markov Model, Support Vector Machine etc. However, a bottleneck to improve performance came into the existence by only using one kind of features. In order to break through it, the fusion approach is introduced into voice biometrics. The objective of Chapter 3 is to show the rationale behind of using fusion methods. At the point of view of biometrics, it systematically classifies the existing approaches into three fusion levels, feature level, matching-score level, and decision-making level. After descriptions of the fundamental basis, each level fusion technique will be described. Then several experimental results will be presented to show the effectiveness of the performance of the fusion techniques.

In Chapter 4 the problem of finding a face recognition system that works well both under variable illumination conditions and under strictly controlled acquisition conditions is considered. The problem under consideration has to do with the fact that systems that work well (compared with standard methods) with variable illumination conditions often suffer a drop in performance on images where illumination is strictly controlled. In this chapter the authors review existing techniques for obtaining illumination robustness and propose a method for handling illumination variance that combines different matchers and preprocessing methods. An extensive evaluation of the authors' system is performed on several datasets (CMU, ORL, Extended YALE-B, and BioLab). The authors' results show that even though some standalone matchers are inconsistent in performance depending on the database, the fusion of different methods performs consistently well across all tested datasets and illumination conditions. The authors' experiments show that the best result are obtained using gradientfaces as a preprocessing method and orthogonal linear graph embedding as a feature transform.

Physical activity (PA) is a modifiable lifestyle factor for many chronic diseases and its health benefits are well known. PA outcomes are often measured and assessed in many clinical and epidemiological studies. Chapter 5 first reviews the problems and issues regarding the analysis of PA outcomes. These include outliers, presence of many zeros and correlated observations, which violate the statistical assumptions and render standard regression analysis inappropriate. An alternative two-part generalized linear mixed models (GLMM) approach is proposed to analyze the heterogeneous and correlated PA data. At the first part, a logistic mixed regression model is fitted to estimate the prevalence of PA and factors associated with PA participation. A gamma mixed regression model is adopted at the second part to assess the effects of predictor variables among those with positive PA outcomes. Variations between clusters are accommodated by random effects within the GLMM framework. The proposed methods are demonstrated using data collected from a community-based study of elderly PA in Japan. The findings provide useful information for targeting physically inactive population subgroups for health promotion programs.

In the last years, numerous research papers have considered the problem of privacy protection in biometric authentication systems (1-to-1). When someone wants to authenticate himself, the fresh biometric data he presents has to match with the reference related to the one he claims to be. Many proposals have thus been made to ensure the confidentiality of the biometric data involved during this verification.

Biometric identification enables to identify an individual among many others without requiring any prior claim of identity (1-to-many). Typically, it consists in checking the belonging of a user to a database. Paradoxically, today, identification constitutes the main application of biometric systems whereas privacy protection in such identification systems has received relatively little attention in the literature. In Chapter 6, the authors show how to extend the previous works on biometric authentication in order to also cover biometric identification while maintaining the privacy of the individuals. This is not a trivial task as the authors want to go much faster than performing comparisons with all the biometric references in the system, i.e. an exhaustive search. The main difficulty comes here from the fact that as they must deal with data that are protected, the authors cannot merely rely on traditional biometric identification methods. This can be overcome successfully thanks to new techniques that have been suggested recently in a few papers.

As discussed in Chapter 7, systematic designs (where the allocation of varieties to plots is according to some fixed, non-random, pattern or scheme) are often used for the controls in large early generation variety trials (EGVTs). These are field trials which compare a large number of new varieties (or lines) with standard/control ones, and which are used to select some high yielding lines for further investigation. Usually the new lines are unreplicated at a site. The efficiency of different types of systematic unreplicated EGVT designs compared to optimal or efficient designs is investigated. Two design optimality criteria are used, and are compared for their association with the probability of selecting the highest yielding varieties. The data are assumed to be spatially dependent, and three models are considered for variety effects: both test and control effects fixed, controls effects fixed but variety effects random; and both test and control variety random.

In: Biometrics: Theory, Applications and Issues
Editor: Ellen R. Nichols, pp. 1-28
ISBN: 978-1-61728-765-7

Chapter 1

SAMPLE SIZE REQUIREMENTS FOR EVALUATING INTERVENTION EFFECTS IN THREE-LEVEL CLUSTER RANDOMIZED CLINICAL TRIALS

Moonseong Heo* and Mimi Y. Kim
Division of Biostatistics,
Department of Epidemiology and Population Health,
Albert Einstein College of Medicine, Bronx, New York

Abstract

Experimental clinical trial settings are now often extended to community entities beyond academic research centers. In such settings, a cluster randomized clinical trial (cluster-RCT) design can be useful to rigorously test the effectiveness of a new intervention. Investigators are most commonly interested in assessing the following three types of intervention effects: overall intervention effect, change in intervention effect over time, and local intervention effect at the end of the study. At the design stage of the cluster-RCT, it is essential to estimate a sample size sufficient for adequate statistical power to evaluate the different intervention effects. However, the sample size estimation must account for the multilevel data structure that is necessitated by the nature of the cluster-RCT design. In this review, we consider a three-level data structure and summarize sample size

* E-mail address: moonseong.heo@einstein.yu.edu and mimi.kim@einstein.yu.edu. (Corresponding author)

approaches for testing intervention effects within a unified framework of mixed-effects linear models which offer flexibility in the analysis of multilevel data and hypotheses testing in a cluster-RCT. The sample size methods are presented in closed form and have been validated by simulation studies. Important features of sample size determination for each primary hypothesis are also discussed.

1. Introduction

A randomized clinical trial in which interventions are assigned at the level of a community entity, such as primary clinics or health organizations, can result in a three-level hierarchical data structure: e.g., subjects are the first level, physicians are the second level, and clinics are the third level in the hierarchy. Such a trial is referred to as a cluster randomized trial [1, 2]. Interventions are often randomly assigned at the highest hierarchical level, the clinic, in order to minimize the contamination bias within clinics that might be observed if random assignment were made at a lower level data unit such as subjects [3, 4]. A three level hierarchical structure can also arise in a longitudinal cluster randomized trial (or longitudinal cluster-RCT) when subjects need to be repeatedly assessed for outcomes during follow-up. In this case, we assume that the subject-level outcome does not depend on physicians within clinics so that the three levels in the hierarchy are the repeated measures (level 1) on each subject, subjects (level 2) within clinics and clinics (level 3).

In three-level cluster randomized trials three types of intervention effects are generally of interest: overall intervention effect, change in intervention effect over time, and local intervention effect at the end of the study. The overall intervention effect is often of primary interest when the outcome is assessed at only a single point in time following treatment in a pre-post study design. Consider a study in which clinics are randomly assigned to either an experimental intervention or usual care for the treatment of depression. Physicians within each clinic treat multiple subjects, who, in turn, are evaluated at a single post-intervention time point for improvement in depression symptoms. In this case the pre-post intervention difference in depressive symptom ratings is the primary outcome and the difference in pre-post symptom improvement between the two interventions can be parameterized as the overall intervention effect of interest.

In longitudinal cluster-RCTs, however, the trajectory of the outcome may differ over time between treatment groups. Therefore, evaluations of the rates of change between intervention groups and the local intervention effect at the end of the study are more relevant. For example, the rate of decline in depression symptoms is expected to be faster in subjects treated with an experimental intervention than with the standard therapy. Under this anticipated trajectory, it would be of greater interest to test differences in the slopes of the depression symptom ratings or to test a local group mean difference in ratings at the end of the study than to test for differences in the overall group-specific means without consideration of time trends.

Regardless of the effect of primary interest and the definition of each level in the hierarchical structure, sample size determination and power calculations are essential in the proper design of a three-level cluster-RCT. The sample sizes or numbers of units at each level that are required to detect a hypothesized effect size with sufficient statistical power must be determined. To this end, we provide explicit closed form power functions and sample size formulae for three-level data in a cluster-RCT to detect each of the three intervention effects, as derived in three previously published papers by the first author [5-7]. The derivations of the power functions and sample size formulae were based on maximum likelihood estimates of the parameters of interest from a mixed-effects linear regression model [8-11] and have been validated in simulation studies.

2. Principles of Statistical Power Calculation

We assume that the outcome variable, Y, is normally distributed with an arbitrary mean and variance, σ^2 and δ is the parameter for the effect of interest. To build a test statistic for evaluating H_0: $\delta = 0$, one has to derive a maximum likelihood estimate, $\hat{\delta}$, and standard error, se($\hat{\delta}$). In general, the maximum likelihood estimate is unbiased, i.e., $E(\hat{\delta}) = \delta$, and the variance of $\hat{\delta}$, i.e., the square of se($\hat{\delta}$), can be expressed as

$$Var\left(\hat{\delta}\right) = \frac{\chi_{1'}\sigma^2}{N},$$

where N is the sample size "per (intervention) group" and $\chi_{1'}$ is a correction factor that depends on study design; e.g., single level or multiple levels; or one group or two groups. For example, if the data structure assumes only one level, $\chi_{1'} = 1$ when the parameter δ represents a population mean in one group, and $\chi_{1'} = 2$ when the parameter δ represents a difference in population means between two groups.

The ratio of $\hat{\delta}$ to se($\hat{\delta}$), that is,

$$D = \frac{\hat{\delta}}{se(\hat{\delta})} = \frac{\hat{\delta}}{\sigma}\sqrt{N/\chi_v}$$

is referred to as the Wald test statistic according to large sample theory [12]. Based on this theory, with known variance σ^2, the test statistic D follows a standard normal distribution, $N(0, 1)$, under the null hypothesis and $D \sim N(\delta/se(\hat{\delta}), 1)$ under the alternative hypothesis of $\delta \neq 0$. Based on this property, a power function φ can be constructed as follows:

$$\varphi = 1 - \beta = \Phi\left\{\frac{\delta}{\sigma}\sqrt{N/\chi_v} - \Phi^{-1}(1-\alpha/2)\right\},$$

where α is a two-sided significance level; β represents the probability of type II error; Φ is the cumulative distribution function (CDF) of a standard normal distribution and Φ^{-1} is its inverse, N is the sample size per group. We make the further assumptions throughout this chapter that: 1) $\delta = |\delta| > 0$; and 2) the probability below a critical value, $\Phi^{-1}(\alpha/2)$, in the other tail is negligible and thus assumed to be 0.

When the parameter of interest is expressed in units of standard deviation of the outcome Y, i.e., in terms of a standardized effect size,

$$\Delta = \delta/\sigma,$$

which is also known as Cohen's d [13, 14], the power function can be simplified as follows:

$$\varphi = \Phi\left\{\Delta\sqrt{\chi_v N} - \Phi^{-1}(1-\alpha/2)\right\}.$$

The sample size can then be determined as

$$N(\Delta) \equiv N(\Delta;\alpha,\beta) = \frac{\chi_v\left\{\Phi^{-1}(1-\alpha/2) + \Phi^{-1}(1-\beta)\right\}^2}{\Delta^2}, \quad (2.1)$$

by taking the inverse of the power function φ to solve for the sample size N. More precisely, N is the smallest integer greater than the right hand side of equation (2.1).

Most importantly, equation (2.1) shows how the sample size N decreases with increasing effect size Δ, and increases with increasing statistical power φ for a given significance level α [15]. This principle applies regardless of the number of levels in the data structure, types of hypotheses, or scale of study outcome, categorical or continuous.

In general, the required sample sizes of a particular level can be expressed in the form of equation (2.1) regardless of the number of levels in the hierarchical data structure. For single level data, required sample sizes *per group* will be N with $\chi_v = 1$ for a one sample or group test and N with $\chi_v = 2$ for a two sample or group test. Sample size determination for two level data has been addressed in the statistical literature and can be expressed in the general form above [e.g., 16-18]. In the following, we will consider three level data and show that the relevant formulas for determining the required sample size of each level in the hierarchy can also be expressed in the same form.

3. Mixed Effects Linear Model for Three-Level Data

A mixed-effects model for comparing two interventions with three-level data can be expressed as follows:

$$Y_{ijk} = \beta_0 + L(X_{ijk}, T_{ijk}) + u_i + u_{j(i)} + e_{ijk} \qquad (3.1)$$

The units in level three are indexed by $i = 1,2,\ldots,2N_3$ (N_3 representing the sample size of the level three units per group), the level two units by $j = 1,\ldots, N_2$ (N_2 representing the sample size of the level two units per level three unit), and the level one units by $k = 1, 2, \ldots, N_1$ (N_1 representing the sample size of the level one units per level two unit). X_{ijk} is the intervention assignment indicator variable equal to 0 if the i-th level three unit is assigned to a control intervention and 1 if assigned to an experimental intervention; therefore $X_{ijk} = X_i$ for all j and k. We assume a balanced design in that $\Sigma_i X_i = N_3$. It is also assumed the time points for subject assessment, T_{ijk}, does not depend on either i or j, i.e., $T_{ijk} = T_k$ for all i and j. L is a linear function of X_{ijk} and T_{ijk} with fixed effect coefficients and depends on the specific hypothesis of interest as described further below.

With respect to the random components, it is assumed that the error term e_{ijk} is normally distributed as $N(0, \sigma_e^2)$, the level two random intercept $u_{j(i)} \sim N(0, \sigma_2^2)$ and the level three random intercept $u_i \sim N(0, \sigma_3^2)$. It is further assumed that

$u_i \perp u_{j(i)} \perp e_{ijk}$, i.e., the three random components are mutually independent. In addition, *conditional independence* is assumed for all $u_{j(i)}$ and for all e_{ijk}, whereas u_i are *unconditionally* independent. That is, $u_{j(i)}$ are independent conditional on u_i, and e_{ijk} are independent conditional on both u_i and $u_{j(i)}$. Under these assumptions, it can be shown that the elements of the covariance matrix are:

$$Cov\left(Y_{ijk}, Y_{i'j'k'}\right) = 1(i = i' \,\&\, j = j' \,\&\, k = k')\sigma_e^2 + 1(i = i' \,\&\, j = j')\sigma_2^2 + 1(i = i')\sigma_3^2, \quad (3.2)$$

where 1(.) is an indicator function. This yields

$$\sigma^2 \equiv Var\left(Y_{ijk}\right) = Cov\left(Y_{ijk}, Y_{ijk}\right) = \sigma_e^2 + \sigma_2^2 + \sigma_3^2 \quad (3.3)$$

Hence, the correlation among level two data can be written for $j \neq j'$ as follows:

$$\rho_2 = Corr\left(Y_{ijk}, Y_{ij'k'}\right) = \frac{\sigma_3^2}{\sigma_e^2 + \sigma_2^2 + \sigma_3^2} = \frac{\sigma_3^2}{\sigma^2} \quad (3.4)$$

The correlation among level one data for $k \neq k'$ can be expressed as:

$$\rho_1 = Corr\left(Y_{ijk}, Y_{ijk'}\right) = \frac{\sigma_2^2 + \sigma_3^2}{\sigma_e^2 + \sigma_2^2 + \sigma_3^2} = \frac{\sigma_2^2 + \sigma_3^2}{\sigma^2} \quad (3.5)$$

4 Hypothesis (I): Overall Main Intervention Effects

4.1. Model Specification

As discussed above, testing the overall intervention effect is of interest when a single post-treatment outcome measurement is obtained for each subject which is the level one unit in this case ($k = 1, 2, ..., N_1$). The subject level outcome can be a difference in severity of depression symptom before and after being treated by a physician using either the experimental or control approach, which was randomly assigned at the clinic level. The physicians who treat multiple subjects are the level two units ($j = 1,2,..,N_2$) and the clinics are the level three unit ($i = 1,2,..,2N_3$). The main or overall effect would represent an intervention effect on pre-post difference in severities of depression symptoms. To determine the required sample sizes to detect the main intervention effect, the L function in the model (3.1) above can be specified as

$$L_{(1)}(X,T) = \delta_{(1)}X_i$$

yielding

$$Y_{ijk} = \beta_0 + \delta_{(1)}X_i + u_i + u_{j(i)} + e_{ijk}. \qquad (4.1)$$

where $\delta_{(1)}$ represents the main intervention effect and β_0 represents a fixed intercept. Accordingly, Hypothesis (I) for the main intervention effect can be expressed as:

$$H_{0(1)} : \delta_{(1)} = 0 \qquad (4.2)$$

The goal is to determine the numbers of clinics per intervention group (N_3), physicians in each clinic (N_2), and subjects treated by each physician (N_1) required to test this hypothesis with a desired statistical power at a pre-specified two-sided significance level.

4.2. Maximum Likelihood Estimates of Hypothesis (I) Parameters

The maximum likelihood estimate (MLE) $\hat{\delta}_{(1)}$ of the overall intervention effect in model (4.1) is the group mean difference in outcome; that is,

$$\hat{\delta}_{(1)} = \bar{Y}_1 - \bar{Y}_0, \qquad (4.3)$$

where

$$\bar{Y}_g = \frac{1}{N_3 N_2 N_1} \sum_{i=1}^{N_3} \sum_{j=1}^{N_2} \sum_{k=1}^{N_1} Y_{ijk}, \qquad (4.4)$$

and $\bar{Y}_g$ ($g = 0,1$) is the group mean of the outcome Y for the g-th group, for which $X_i = g$. The MLE $\hat{\delta}$ (4.3) is indeed the ordinary least squares estimate of the group mean difference, if N_1 is fixed and does not vary across the second level units [9]. It is unbiased since

$$E\left(\hat{\delta}_{(1)}\right)=E\left(\bar{Y}_1\right)-E\left(\bar{Y}_0\right)=(\beta_0+\delta_{(1)})-\beta_0=\delta_{(1)}.$$

Further, the variance of group mean $\bar{Y}_g$ can be obtained based on equation (3.2) as follows:

$$Var\left(\bar{Y}_g\right)=\frac{1}{N_3N_2N_1}\left\{\sigma^2+N_1(N_2-1)\sigma_3^2+(N_1-1)(\sigma_3^2+\sigma_2^2)\right\}.$$

This variance can be rewritten as a function of two correlations (3.4) and (3.5) as follows:

$$Var\left(\bar{Y}_g\right)=\frac{\sigma^2}{N_3N_2N_1}\left\{1+N_1(N_2-1)\rho_2+(N_1-1)\rho_1\right\}.$$

Thus, compared to a single level data structure with $\rho_1=\rho_2=0$ or equivalently $\sigma_2^2=\sigma_3^2=0$, the variance of a group mean $\bar{Y}_g$ under a three level data structure is inflated by a *variance inflation factor* or *design effect* denoted by f, i.e.,

$$Var\left(\bar{Y}_g\right)=\frac{f\sigma^2}{N_3N_2N_1}, \tag{4.5}$$

where

$$f=1+N_1(N_2-1)\rho_2+(N_1-1)\rho_1, \tag{4.6}$$

which is not dependent on N_3. In contrast, if $\rho_1=\rho_2=0$, then $f=1$ and $Var(\bar{Y}_g)=\sigma^2/(N_3N_2N_1)$ for each g.

4.3. Power and Sample Size for Testing Hypothesis (I)

The following test statistic $D_{(I)}$, based on (4.3) and (4.5), can be used to test the null hypothesis (4.2):

$$D_{(I)} = \frac{\hat{\delta}_{(I)}}{se(\hat{\delta}_{(I)})} = \frac{\hat{\delta}_{(I)}}{\sqrt{Var\left(\bar{Y}_1\right) + Var\left(\bar{Y}_0\right)}} = \frac{\sqrt{N_3 N_2 N_1}\left(\bar{Y}_1 - \bar{Y}_0\right)}{\sigma\sqrt{2f}}.$$

The power of the test statistic $D_{(I)}$, denoted by $\varphi_{(I)}$, can therefore be written as follows:

$$\varphi_{(I)} = 1 - \beta = \Phi\left\{\Delta_{(I)}\sqrt{\frac{N_3 N_2 N_1}{2f}} - \Phi^{-1}(1-\alpha/2)\right\}, \quad (4.7)$$

where $\Delta_{(I)} = \delta_{(I)} / \sigma$, i.e., Cohen's d or the standardized effect size, is the mean difference in the outcome Y expressed in units of a pooled within-group standard deviation (SD) σ, the square root of equation (3.3).

It follows that when the hypothesis test is based on $D_{(I)}$ with a two-sided significance level of α, the required level three sample size per group, N_3, to attain statistical power $\varphi_{(I)} = 1 - \beta$ can be calculated from equation (4.7) as:

$$N_3\left(\Delta_{(I)}\right) = \frac{2f\left\{\Phi^{-1}(1-\alpha/2) + \Phi^{-1}(1-\beta)\right\}^2}{N_2 N_1 \Delta_{(I)}^2}. \quad (4.8)$$

It can be seen that the effect of ρ_2 on the sample size N_3 through f is greater than that of ρ_1 because $\partial f/\partial \rho_2 = N_1(N_2 - 1) > \partial f/\partial \rho_1 = (N_1 - 1)$. At the same time, the effect of σ_3 is greater than that of σ_2 based on (4.5) and (4.6). Comparing (2.1) and (4.8), we have

$$\chi_{1'}\left(\Delta_{(I)}\right) = \frac{2f^2}{N_2 N_1},$$

Therefore $N_3(\Delta_{(I)})$ is a special form of equation (2.1). The sample sizes N_2 and N_1 for level 2 and 1 units, respectively, can be obtained by solving equation (4.7) for each of them as follows:

$$N_2\left(\Delta_{(I)}\right)=\frac{2\{1+(\rho_1-\rho_2)N_1-\rho_1\}\{\Phi^{-1}(1-\alpha/2)+\Phi^{-1}(1-\beta)\}^2}{N_1N_3\Delta_{(I)}^2-2\rho_2N_1\{\Phi^{-1}(1-\alpha/2)+\Phi^{-1}(1-\beta)\}^2},$$

and

$$N_1\left(\Delta_{(I)}\right)=\frac{2(1-\rho_1)\{\Phi^{-1}(1-\alpha/2)+\Phi^{-1}(1-\beta)\}^2}{N_2N_3\Delta_{(I)}^2-2\{(N_2-1)\rho_2+\rho_1\}\{\Phi^{-1}(1-\alpha/2)+\Phi^{-1}(1-\beta)\}^2}.$$

5. Hypothesis (II): Time-by-Interaction Effects

5.1. Model Specification

Longitudinal cluster-RCT designs aiming to test differences in trends in outcome over time, such as decline in depression symptom severity between intervention groups, naturally entails repeated outcome assessments at specific time points. These repeated measures comprise the level one unit ($k = 1, 2, ..., N_1$). The repeated measures are nested or clustered within study subjects, the level two units (j=1,2,..,N_2), who are in turn nested within clinics, the level three units (i = 1,2, ..,2N_3). Each clinic is randomly assigned to administer an experimental or control therapy to all subjects seen by physicians within the clinic. Trends in the outcome over time can be compared between the experimental and control groups by including an interaction effect, $\delta_{(II)}$, between treatment group and time in the L function in model (3.1) as follows:

$$L_{(II)}(X,T)=\xi X_i+\tau T_k+\delta_{(II)}X_iT_k.$$

This yields

$$Y_{ijk}=\beta_0+\xi X_i+\tau T_k+\delta_{(II)}X_iT_k+u_i+u_{j(i)}+e_{ijk}, \quad (5.1)$$

where $\delta_{(II)}$ represents the parameter of primary interest, the time-by-intervention effect that quantifies the slope difference in outcome Y between the two intervention groups, or additional decline of symptom severity in the experimental group compared to the control group. The parameter ξ represents the intervention effect at baseline, i.e., when T_k = 0, the parameter τ represents the decline in symptom severity over time in the control group, and the decline in symptom severity over time in the intervention group is $\tau + \delta_{(II)}$. The overall fixed intercept is denoted by β_0. Accordingly, Hypothesis (II) can be written as:

$$H_{0(II)} : \delta_{(II)} = 0 , \quad (5.2)$$

To test this hypothesis with sufficient power, the required numbers of clinics per intervention group (N_3), the number subjects treated by each clinic (N_2) and the number of repeated assessments on each subject (N_1) need to be determined.

5.2. Maximum Likelihood Estimates of Hypothesis (II) Parameters

The maximum likelihood estimate $\hat{\delta}_{(II)}$ of the interaction effect in model (5.1) is indeed the slope difference between the two groups: that is,

$$\hat{\delta}_{(II)} = \hat{\eta}_1 - \hat{\eta}_0 , \quad (5.3)$$

where $\hat{\eta}_g$ ($g = 0,1$) is the MLE of the slope for the outcome Y in the g-th group, in which $X_i = g$. Specifically, for i in the g-th group,

$$\begin{aligned}\hat{\eta}_g &= \sum_{i=1}^{N_3}\sum_{j=1}^{N_2}\sum_{k=1}^{N_1}(T_k - \bar{T})(Y_{ijk} - \bar{Y}_g) \Big/ \sum_{i=1}^{N_3}\sum_{j=1}^{N_2}\sum_{k=1}^{N_1}(T_k - \bar{T})^2 \\ &= \sum_{i=1}^{N_3}\sum_{j=1}^{N_2}\sum_{k=1}^{N_1}(T_k - \bar{T})(Y_{ijk} - \bar{Y}_g) \Big/ N_3N_2N_1Var_p(T)\end{aligned} , \quad (5.4)$$

is the ordinary least squares estimate of the slope with fixed non-varying N_1 [9], where: 1) $\bar{Y}_g$ ($g = 0,1$) is the overall group mean of the outcome Y for the g-th group; 2) $\bar{T} = \sum_{k=1}^{N_1} T_k / N_1$ is the "mean" time point; and 3) $Var_p(T) = \sum_{k=1}^{N_1}(T_k - \bar{T})^2 / N_1$ is the "population variance" of the time variable T.

It can be shown that the MLE $\hat{\delta}_{(II)}$ (5.3) is unbiased, i.e., $E\left(\hat{\delta}_{(II)}\right)=E\left(\hat{\eta}_1-\hat{\eta}_0\right) = \left(\tau+\delta_{(II)}\right)-\tau=\delta_{(II)}$. The variance of a slope MLE $\hat{\eta}_g$ can be obtained based on equation (3.2) as follows (see [6] for a proof):

$$Var\left(\hat{\eta}_g\right)=\frac{\sigma_e^2}{N_3N_2N_1Var_p(T)}=\frac{(1-\rho_1)\sigma^2}{N_3N_2N_1Var_p(T)}.$$

Therefore, the variance of $\hat{\delta}_{(II)}$ is

$$Var\left(\hat{\delta}_{(II)}\right)=Var\left(\hat{\eta}_1-\hat{\eta}_0\right)=Var\left(\hat{\eta}_1\right)+Var\left(\hat{\eta}_0\right)=\frac{2(1-\rho_1)\sigma^2}{N_3N_2N_1Var_p(T)}. \quad (5.5)$$

Observe that $\hat{\eta}_1$ and $\hat{\eta}_0$ are independent of each other. It is notable, however, that the variance of $\hat{\delta}_{(II)}$ depends only on the residual variance σ_e^2, and not on σ_3^2, σ_2^2, or ρ_2. Therefore, *for a given total variance* σ^2, it decreases with decreasing σ_e^2 or increasing ρ_1, the correlation among the first level data.

5.3. Power and Sample Size for Testing Hypothesis (II)

The following test statistic $D_{(II)}$, based on (5.3) and (5.5), can be used to test the null hypothesis (5.2):

$$D_{(II)}=\frac{\hat{\delta}_{(II)}}{se(\hat{\delta}_{(II)})}=\frac{\hat{\delta}_{(II)}}{\sqrt{Var(\hat{\eta}_1)+Var(\hat{\eta}_0)}}=\frac{\sqrt{N_3N_2N_1Var_p(T)}\left(\hat{\eta}_1-\hat{\eta}_0\right)}{\sigma\sqrt{2(1-\rho_1)}}.$$

The power of the test statistic $D_{(II)}$, denoted by $\varphi_{(II)}$, can accordingly be written as follows:

$$\varphi_{(II)}=1-\beta=\Phi\left[\Delta_{(II)}\sqrt{\frac{N_3N_2N_1Var_p(T)}{2(1-\rho_1)}}-\Phi^{-1}(1-\alpha/2)\right], \quad (5.6)$$

where $\Delta_{(II)} = \delta_{(II)} /\sigma$, i.e., the slope difference is expressed in SD units, that is, a standardized effect size of the interaction.

It follows that the required third level unit sample size N_3 per group for a desired statistical power $\varphi_{(II)} = 1 - \beta$ can be calculated from equation (5.6) as:

$$N_3\left(\Delta_{(II)}\right) = \frac{2\left(\Phi^{-1}(1-\alpha/2)+\Phi^{-1}(1-\beta)\right)^2(1-\rho_1)}{N_2N_1Var_p(T)\Delta_{(II)}^2} \quad (5.7)$$

It can be observed that the level 3 sample size is not a function ρ_2 and that it is a decreasing function of increasing ρ_1 and $Var_p(T)$ in particular. Therefore, more follow-up with more consistent (as opposed to erratic) observations within subjects over time will increase the power (5.6) and at the same time will reduce the required N_3 or N_2 for the same anticipated power. Comparing (2.1) and (5.7), we have

$$\chi_{1'}\left(\Delta_{(II)}\right) = \frac{2(1-\rho_1)}{N_2N_1Var_p(T)}$$

Therefore, $N_3(\Delta_{(II)})$ is also a special form of equation (2.1). The sample size N_2 has a reciprocal relationship with N_3 in the sense that the power depends on N_2N_3 because both are independent of each other. Therefore, sample size N_2 for the level two data can immediately be determined from equation (5.7) as follows:

$$N_2\left(\Delta_{(II)}\right) = \frac{2\left(\Phi^{-1}(1-\alpha/2)+\Phi^{-1}(1-\beta)\right)^2(1-\rho_1)}{N_3N_1Var_p(T)\Delta_{(II)}^2}.$$

The sample size N_1 for the level one data should, however, be determined in an iterative manner because $Var_p(T)$ is a function of N_1. Specifically, an iterative solution for N_1 must satisfy the following equation:

$$N_1\left(\Delta_{(II)}\right) = \frac{2\left(\Phi^{-1}(1-\alpha/2)+\Phi^{-1}(1-\beta)\right)^2(1-\rho_1)}{N_3N_2Var_p(T)\Delta_{(II)}^2}.$$

6. Hypothesis (III): Intervention Effects at the End of Trial

6.1. Model Specification

In a longitudinal cluster-RCT, it is often anticipated as in Hypothesis (II) above that the intervention effect is expected to be gradual, resulting in a decline in symptom severity which will diverge over time between subjects treated with different interventions. Under this anticipated trajectory, it may be of greater interest to test a *local* group mean difference at the end of the study than to test for differences in the overall group-specific means. Again, the repeated assessments on each subject are the level one unit (k = 1, 2, ...,N_1), the subjects nested within clinics are the level two units (j=1,2,..,N_2), and the clinics are the level three units (i = 1,2, ..,2N_3). To parameterize the local group mean difference, we first assume that the time variable T_k increases from 0 (the baseline) to $T_{\text{end}} = N_1 - 1$ (the last time point) by 1 with equal time intervals. Then a contrast representing the local intervention effect can be constructed based on a shifted scale of the time variable $T_k' = T_k - T_{\text{end}}$ and the L function can be specified as follows:

$$L_{(\text{III})}(X,T) = \delta_{(\text{III})}X_i + \tau(T_k - T_{end}) + \gamma X_i(T_k - T_{end})$$

yielding

$$Y_{ijk} = \beta_0 + \delta_{(\text{III})}X_i + \tau(T_k - T_{end}) + \gamma X_i(T_k - T_{end}) + u_i + u_{j(i)} + e_{ijk} \,. \quad (6.1)$$

The parameter $\delta_{(\text{III})}$ represents the intervention effect at the end of the study. The parameter τ represents the slope for the time effect in the control group, that is, the decline in symptom severity over time in the control group. The parameter γ represents the intervention by time effect, which corresponds to the slope difference in outcome Y between the intervention groups or the additional decline in the experimental group relative to the control group. The overall fixed intercept is denoted by β_0 on the shifted time scale $T_k' = T_k - T_{\text{end}}$. Accordingly, Hypothesis (III) can be written as:

$$\text{H}_{0(\text{III})} : \delta_{(\text{III})} = 0 \,. \quad (6.2)$$

To test this hypothesis with sufficient power, the required numbers of clinics per intervention group (N_3), the number subjects treated by each clinic (N_2) and the number of repeated assessments on each subject (N_1) need to be determined as in Hypothesis (II) (5.2) above.

6.2. Maximum Likelihood Estimates of Hypothesis (III) Parameters

By virtue of random assignments of the two interventions, we assume no mean difference in outcome Y between the two groups at baseline on the original time scale T. Under this assumption, the maximum likelihood estimate $\hat{\delta}_{(III)}$ in model (6.1) can be obtained as:

$$\hat{\delta}_{(III)} = \hat{\theta} - \hat{\gamma}\overline{T}' = \hat{\theta} + \hat{\gamma}\overline{T}, \qquad (6.3)$$

where: 1) $\overline{T}' = \sum_{k=1}^{N_1} T_k'/N_1$ and $\overline{T} = \sum_{k=1}^{N_1} T_k/N_1$ are the "means" of the shifted and original time point, respectively (note that $\overline{T} = -\overline{T}'$); 2) $\hat{\theta} = \overline{Y}_1 - \overline{Y}_0$ is the overall group mean difference, where $\overline{Y}_g$ is defined in (4.4) is the overall group mean of the outcome Y for the g-th group, for which $X_i = g$; and 3) $\hat{\gamma} = \hat{\eta}_1 - \hat{\eta}_0$ is the slope difference between the two groups in (5.4). In fact, the slope estimates are invariant over choice between the shifted time variable T_k' and the original counterpart T_k. Again, the ordinary least squares estimates of $\hat{\theta} = \overline{Y}_1 - \overline{Y}_0$ and $\hat{\gamma} = \hat{\eta}_1 - \hat{\eta}_0$ are maximum likelihood estimates under the assumption of a perfectly balanced design, i.e., N_1 does not vary across the level two data units [9]. Therefore, $\hat{\delta}_{(III)}$ (6.3) is the maximum likelihood estimate.

The MLE $\hat{\delta}_{(III)}$ (6.3) is unbiased: $E(\hat{\theta}) = E(\overline{Y}_1 - \overline{Y}_0) = \beta_0' + \delta_{(III)} + \tau\overline{T}' + \gamma\overline{T}' - (\beta_0' + \tau\overline{T}') = \hat{\delta}_{(III)} + \gamma\overline{T}'$; and $E(\hat{\gamma}) = E(\hat{\eta}_1 - \hat{\eta}_0) = \eta_1 - \eta_0 = \gamma$. It follows that $E(\hat{\delta}_{(III)}) = E(\hat{\theta}) - E(\hat{\gamma})\overline{T}' = \delta_{(III)} + \gamma\overline{T}' - \gamma\overline{T}' = \delta_{(III)}$. Since the "mean" and "slope" estimates are independent, we obtained the variance of $\hat{\delta}_{(III)}$ as follows (see [7] for details):

$$Var(\hat{\delta}_{(III)}) = Var(\hat{\theta}) + \overline{T}'^2 Var(\hat{\gamma}) = \frac{2f\sigma^2}{N_3 N_2 N_1} C_f, \qquad (6.4)$$

where f is the variance inflation factor or design effect (4.6) and

$$C_f = \left\{1 + (1-\rho_1)CV^{-2}(T')/f\right\} . \quad (6.5)$$

This is a correction factor to the variance $Var\left(\hat{\theta}\right) = 2f\sigma^2/(N_3N_2N_1)$, where $CV(T') = SD_p(T')/\bar{T}'$ is the coefficient of variation (CV) of the time variable T', and $SD_p(T')$ is a "population" standard deviation, the square root of $Var_p(T')$. Finally, it should be noted that $Var\left(\hat{\delta}_{(III)}\right) \neq 4Var\left(\hat{\theta}\right)$ despite the fact that $\hat{\delta}_{(III)} = 2\hat{\theta}$. At the same time, $Var\left(\delta_{(III)}\right) \neq T_{end}^2 Var\left(\hat{\gamma}\right)$, again despite $\hat{\delta}_{(III)} = T_{end}\hat{\gamma}$.

6.3. Power and Sample Size for Testing Hypothesis (III)

The following test statistic $D_{(III)}$, based on (6.3) and (6.4), can be used to test the null hypothesis (6.2):

$$D_{(III)} = \frac{\hat{\delta}_{(III)}}{se(\hat{\delta}_{(III)})} = \frac{\hat{\delta}_{(III)}}{\sqrt{Var(\hat{\delta}_{(III)})}} = \frac{\sqrt{N_3N_2N_1}\hat{\delta}_{(III)}}{\sigma\sqrt{2fC_f}} .$$

When the difference in group means at the end of a study is expressed in units of the pooled within-group standard deviation, i.e., in terms of a standardized effect size, $\Delta_{(III)} = \delta_{(III)}/\sigma$, the power function can be expressed as follows:

$$\varphi_{(III)} = \Phi\left\{\Delta_{(III)}\sqrt{\frac{N_3N_2N_1}{2fC_f}} - \Phi^{-1}(1-\alpha/2)\right\} . \quad (6.6)$$

It follows that when hypothesis testing is based on $D_{(III)}$ with a two-sided significance level of α, the third level unit sample size N_3 per group for a desired statistical power $\varphi_{(III)} = 1 - \beta$ can be calculated from equation (6.6) as:

$$N_3\left(\Delta_{(III)}\right) = \frac{2fC_f\left(\Phi^{-1}(1-\alpha/2) + \Phi^{-1}(1-\beta)\right)^2}{N_2N_1\Delta_{(III)}^2} . \quad (6.7)$$

The equation (6.7) leads to

$$\chi_{1'}\left(\Delta_{(III)}\right)=\frac{2fC_f}{N_2N_1}$$

a special form of equation (2.1). When compared with the sample size required to detect a hypothesized group mean difference using only data from the end of the trial T_{end}, the sample size $N_3(\Delta_{(III)})$ is smaller than that, say $\hat{N}_3$, derived under two level data structure. Specifically, utilization of only data from the end of the trial will reduce the three level data structure to a two level structure because the other repeated measurements are ignored, i.e., $N_1 = 1$. Then the required sample size (e.g., number of clinics) $\hat{N}_3$ with N_2 (e.g., number of subjects) based on a two level-data structure can be written as follows [17]:

$$\hat{N}_3=\frac{2(1+(N_2-1)\rho_2)\left(\Phi^{-1}(1-\alpha/2)+\Phi^{-1}(1-\beta)\right)^2}{N_2\Delta_\delta^2}.$$

This sample size is indeed identical to $N_3(\Delta_{(III)})$ when N_1 is replaced by 1 (assuming that $CV^{-2}(T')$ is 0 in this case) as one would expect. Furthermore, under two it serves as an upper bound for $N_3(\Delta_{(II)})$ regardless of N_1 and ρ_1. Therefore, when a longitudinal cluster-RCT design is implemented, incorporation of all outcome measurements increases the statistical power to test a group mean difference at the end of the trial.

The sample size N_2 of the level data units can be obtained by solving for it from equation (6.6) as follows:

$$N_2\left(\Delta_{(III)}\right)=\frac{2\left\{1-\rho_1+(\rho_1-\rho_2)N_1+(1-\rho_1)CV^{-2}(T')\right\}\left\{\Phi^{-1}(1-\alpha/2)+\Phi^{-1}(1-\beta)\right\}^2}{N_1N_3\Delta_{(III)}^2-2\rho_2N_1\left\{\Phi^{-1}(1-\alpha/2)+\Phi^{-1}(1-\beta)\right\}^2}. \quad (6.8)$$

The sample size N_1 for the level one data should, however, be determined in an iterative manner because $CV(T')=SD_p(T')/\bar{T}'$ is a function of N_1. The iterative solution for N_1 must satisfy the following equation:

$$N_1\left(\Delta_{(III)}\right)=\frac{2\left\{1-\rho_1+(1-\rho_1)CV^{-2}(T')\right\}\left\{\Phi^{-1}(1-\alpha/2)+\Phi^{-1}(1-\beta)\right\}^2}{N_2N_3\Delta^2_{(III)}-2\left\{(N_2-1)\rho_2+\rho_1\right\}\left\{\Phi^{-1}(1-\alpha/2)+\Phi^{-1}(1-\beta)\right\}^2}. \quad (6.9)$$

Since the CV of the time variable T (or T') is invariant over both location and scale changes, elongation of the time intervals will not affect the sample size (6.7)-(6.9). For example, required sample sizes with time intervals $t = T_k - T_{k-1}$ for all k will be the same as those with time intervals ωt for any $\omega > 0$.

7. Summary and Applications

7.1. Sample Size Calculations for Testing Hypothesis (I)

Table 1 summarizes estimated sample sizes $N_3(\Delta_{(I)})$ based on equation (4.8) required to detect the standardized main intervention effect $\Delta_{(I)}$ with given statistical power = 80%, ρ_1, ρ_2, N_2 and N_1 at a two-side significance level α=0.05. The effect of the standardized effect size $\Delta_{(I)}$ on sample size N_3 is the greatest. For example, when N_2 = 5, the sample size N_3 ranges: from 19 to 40 for $\Delta_{(I)}$ = 0.3; from 11 to 23 for $\Delta_{(I)}$ = 0.4; from 7 to 15 for $\Delta_{(I)}$ = 0.5. Indeed, for $\Delta_{(I)}$ as small as 0.5, the effect of the other parameters on sample size are relatively small because variation in $N_3(\Delta_{(I)})$ is small. The effect of ρ_2 on sample size N_3 is shown to be much greater than that of ρ_1 as expected for the reasons described above in section 4.3. At the same time, the effect of increases in N_2 on the reduction in N_3 is much greater than the effect of increases in N_1. Nevertheless, the effect of N_2 may not be substantial particularly for larger ρ_2 as it can be observed that increasing N_2 from 25 to 50 would reduce $N_3(\Delta_{(I)})$ at most by 2 (Table 1) for 80% power, keeping the other identical design parameters fixed.

Sample sizes presented in Table 1 can be used to design a cluster-RCT that compares subjects treated by physicians in clinics randomly assigned to an experimental intervention with those treated by physicians in clinics assigned to a control intervention. Suppose that for research purposes each clinic can accommodate 5 physicians (N_2), each of whom would be able to provide care to 6 depressed subjects (N_1). The results in Table 1 can be applied to estimating the number of clinics, i.e., level 3 units (N_3), for 80% power to detect various effect sizes (Δ). If ρ_1 = 0.05 and ρ_2 = 0.5, then 31 clinics (N_3) per group, or a total of 62 clinics, would be needed to detect an effect size $\Delta_{(I)}$ = 0.3 with at least 80% statistical power (Table 1).

7.2. Sample Size Calculations for Testing Hypothesis (II)

Table 2 summarizes estimated sample sizes $N_3(\Delta_{(II)})$ based on equation (5.7) required to detect the time-by-intervention interaction effect $\Delta_{(II)}$ with given statistical power = 80%, ρ_1, N_2 and N_1 at a two-side significance level α=0.05. Recall that $N_3(\Delta_{(II)})$ is not a function of ρ_2. However, the effect size of the interaction, $\Delta_{(II)}$, is specified in a way that would yield a standardized between-group mean difference, $\Delta_{(II)}T_{end}$, at the end of the trial, i.e., when $T = T_{end} = N_1 - 1$. That is, the difference $\Delta_{(II)}T_{end}$ serves as a reference value to determine the standardized slope difference $\Delta_{(II)}$. As expected, the sample size $N_3(\Delta_{(II)})$ decreases with increasing correlation ρ_1 for fixed power and other design parameters held constant. For example, when $N_2 = 5$, $N_1 = 6$, and $\Delta_{(II)}T_{end} = 0.3$, (or $\Delta_{(II)} = 0.3/5 = 0.06$) the respective sample sizes requirements for 80% power for the level three data were $N_3(\Delta_{(II)}) = 30$, 25, and 20 for $\rho_1 = 0.4$, 0.5, and 0.6. Furthermore, the power is identical for various combinations of N_2 and N_3 which yield the same product N_2N_3 assuming other design parameters are held constant. When $N_1 = 3$, $\rho_1 = 0.4$, and $\Delta_{(II)}T_{end} = 0.3$ (or $\Delta_{(II)} = 0.3/2 = 0.15$), the product of each the following pairs of N_2 and N_3 is 210 for 80% power: $N_2 = 5$ and $N_3 = 42$; $N_2 = 10$ and $N_3 = 21$; $N_2 = 30$ and $N_3 = 7$ (Table 2).

The results in Table 2 can be applied to design a longitudinal cluster-RCT. Consider a longitudinal cluster-RCT that compares an innovative primary care level intervention with usual primary care practice for the treatment of depression, as conducted in the PROSPECT [19, 20] and the RESPECT [21] trials. It is assumed that each primary clinic can enroll 20 subjects (N_2) for this trial and each subject would be followed up 6 times (N_1) for clinical assessments. To have adequate power for testing whether the course of depressive symptoms over time depends on the specific care that the subjects receive, the results in Table 2 can be used to estimate the required number of primary clinics, i.e., level 3 units (N_3). If $\rho_1 = 0.5$, then four clinics (N_3) for each of the two intervention groups, or a total of 160 subjects, would be needed to detect an effect size $\Delta_{(II)}T_{end} = 5\Delta_{(II)} = 0.4$ (or $\Delta_{(II)} = 0.4/5 = 0.08$) with at least 80% statistical power.

Table 1. Sample size $N_3(\Delta_{(1)})$ to detect a main intervention effect with 80% statistical power at a two-sided significance level $\alpha = 0.05$ under combination of N_1, N_2, ρ_1, ρ_2 and $\Delta_{(1)}$

			$\Delta_{(1)} = 0.3$			$\Delta_{(1)} = 0.4$			$\Delta_{(1)} = 0.5$		
N_2	N_1	ρ_2	ρ_1=0.4	ρ_1=0.5	ρ_1=0.6	ρ_1=0.4	ρ_1=0.5	ρ_1=0.6	ρ_1=0.4	ρ_1=0.5	ρ_1=0.6
5	3	0.01	23	25	27	13	14	16	9	9	10
		0.05	28	31	33	16	18	19	11	11	12
		0.10	35	38	40	20	21	23	13	14	15
	6	0.01	19	22	25	11	13	14	7	8	9
		0.05	25	28	31	14	16	18	9	10	11
		0.10	32	35	38	18	20	21	12	13	14
10	3	0.01	13	14	15	7	8	9	5	5	6
		0.05	19	20	21	11	11	12	7	8	8
		0.10	27	28	29	15	16	17	10	10	11
	6	0.01	11	12	14	6	7	8	4	5	5
		0.05	17	19	20	10	11	11	6	7	8
		0.10	25	26	28	14	15	16	9	10	10
25	3	0.01	6	7	7	4	4	4	3	3	3
		0.05	13	14	14	8	8	8	5	5	5
		0.10	21	22	22	12	13	13	8	8	8
	6	0.01	6	6	7	3	4	4	2	3	3
		0.05	12	13	14	7	7	8	5	5	5
		0.10	21	21	22	12	12	13	8	8	8

Table 1. Continued

			$\Delta_{(I)} = 0.3$			$\Delta_{(I)} = 0.4$			$\Delta_{(I)} = 0.5$		
N_2	N_1	ρ_2	ρ_1=0.4	ρ_1=0.5	ρ_1=0.6	ρ_1=0.4	ρ_1=0.5	ρ_1=0.6	ρ_1=0.4	ρ_1=0.5	ρ_1=0.6
50	3	0.01	4	5	5	3	3	3	2	2	2
		0.05	11	11	12	6	7	7	4	4	4
		0.10	20	20	20	11	11	12	7	7	8
	6	0.01	4	4	5	2	3	3	2	2	2
		0.05	11	11	11	6	6	7	4	4	4
		0.10	19	20	20	11	11	11	7	7	7

N_1 = the number of level one units; N_2 = the number of level two units; ρ_1 = correlation among level one data; ρ_2 = correlation among level two data; $\Delta(I)$ = standardized effect size of the intervention.

Table 2. Sample size $N_3(\Delta_{(II)})$ to detect a time-by-intervention interaction effect with 80% statistical power at a two-sided significance level $\alpha = 0.05$ under combination of N_1, N_2, ρ_1, and $\Delta_{(II)}$

		$\Delta_{(II)}T_{end} = 0.3$			$\Delta_{(II)}T_{end} = 0.4$			$\Delta_{(II)}T_{end} = 0.5$		
N_2	N_1	ρ_1=0.4	ρ_1=0.5	ρ_1=0.6	ρ_1=0.4	ρ_1=0.5	ρ_1=0.6	ρ_1=0.4	ρ_1=0.5	ρ_1=0.6
5	3	42	35	28	24	20	16	16	13	11
	6	30	25	20	17	15	12	11	9	8
	12	18	15	12	10	9	7	7	6	5
10	3	21	18	14	12	10	8	8	7	6
	6	15	13	10	9	8	6	6	5	4
	12	9	8	6	5	5	4	4	3	3

Table 2. Continued

		$\Delta_{(II)}T_{end} = 0.3$			$\Delta_{(II)}T_{end} = 0.4$			$\Delta_{(II)}T_{end} = 0.5$		
N_2	N_1	ρ_1=0.4	ρ_1=0.5	ρ_1=0.6	ρ_1=0.4	ρ_1=0.5	ρ_1=0.6	ρ_1=0.4	ρ_1=0.5	ρ_1=0.6
20	3	11	9	7	6	5	4	4	4	3
	6	8	8	7	5	4	3	3	3	2
	12	5	4	3	3	3	2	2	2	2
30	3	7	6	5	4	4	3	3	3	2
	6	5	5	4	3	3	2	2	2	2
	12	3	3	2	2	2	2	2	1	1

N_1 = the number of level one units; N_2 = the number of level two units; ρ_1 = correlation among level one data; $\Delta_{(II)}$ = standardized effect size of the slope difference that yields an intervention effect $\Delta_{(II)}T_{end}$ at the end of a study.

Table 3. Sample size $N_3(\Delta_{(III)})$ to detect a time-by-intervention interaction effect with 80% statistical power at a two-sided significance level $\alpha = 0.05$ under combination of N_1, N_2, ρ_1, ρ_2 and $\Delta_{(II)}$

			$\Delta_{(III)}$= 0.3			$\Delta_{(III)}$= 0.4			$\Delta_{(III)}$= 0.5		
N_2	N_1	ρ_2	ρ_1=0.3	ρ_1=0.5	ρ_1=0.7	ρ_1=0.3	ρ_1=0.5	ρ_1=0.7	ρ_1=0.3	ρ_1=0.5	ρ_1=0.7
5	5	0.05	33	35	38	19	20	22	12	13	14
		0.10	40	42	45	22	24	26	15	16	17
	9	0.05	27	32	36	16	18	20	10	12	13
		0.10	34	38	43	19	22	24	13	14	16
	13	0.05	25	30	35	14	17	20	9	11	13
		0.10	32	37	42	18	21	24	12	14	15

Table 3. Continued

			$\Delta_{(III)}$= 0.3			$\Delta_{(III)}$= 0.4			$\Delta_{(III)}$= 0.5		
N_2	N_1	ρ_2	ρ_1=0.3	ρ_1=0.5	ρ_1=0.7	ρ_1=0.3	ρ_1=0.5	ρ_1=0.7	ρ_1=0.3	ρ_1=0.5	ρ_1=0.7
10	5	0.05	21	22	24	12	13	14	8	8	9
		0.10	30	30	32	16	17	18	11	11	12
	9	0.05	18	20	23	10	12	13	7	8	8
		0.10	26	28	30	15	16	17	10	10	11
	13	0.05	17	19	22	10	11	13	6	7	8
		0.10	25	27	30	14	16	17	9	10	11
20	5	0.05	15	16	16	9	9	9	6	6	6
		0.10	23	24	25	13	14	14	9	9	9
	9	0.05	14	15	16	8	9	9	5	6	6
		0.10	22	23	24	13	13	14	8	9	9
	13	0.05	13	14	16	8	8	9	5	5	6
		0.10	21	23	24	12	13	14	8	8	9

N_1 = the number of level one units; N_2 = the number of level two units; ρ_1 = correlation among level one data; ρ_2 = correlation among level two data; $\Delta_{(III)}$ = standardized effect size of the intervention effect at the end of study.

7.3. Sample Size Calculations for Testing Hypothesis (III)

Table 3 summarizes estimated sample sizes $N_3(\Delta_{(III)})$ based on equation (6.7) required to detect the local intervention effect at the end of study $\Delta_{(III)}$ with given statistical power = 80%, ρ_1, ρ_2, N_2 and N_1 at a two-sided significance level α=0.05. One important observation is as follows. The variance of the group mean difference at the end of a study under model (6.1) is a combination of the variance of the sample overall means and that of the estimated slope as shown in equation (6.4). The former variance increases with increasing ρ_1 (4.5) whereas the latter decreases with increasing ρ_1 (5.5). Nevertheless, it can be seen in table 3 that the required sample sizes increase with increasing ρ_1 because the positive effect of ρ_1 on the $Var(\hat{\theta})$ is greater than its negative effect on $Var(\hat{\gamma})$ in equation (6.4).

The results presented in Table 3 can be applied to estimate the number of clinics, i.e., level 3 units (N_3), for 80% power to detect various local effect sizes $\Delta_{(III)}$. For example, suppose that each available clinic can recruit 10 subjects with depression (N_2), each of whom would be assessed 9 times (N_1) including a baseline assessment. If $\rho_1 = 0.05$ and $\rho_2 = 0.5$, then 12 clinics (N_3) per group, or a total of 24 clinics, would be needed to detect an effect size $\Delta_{(III)} = 0.4$ with at least 80% statistical power. If ρ_1 is assumed to be 0.10, then 16 clinics (N_3) per group, or a total of 32 clinics, would be needed (Table 3).

8. Discussion

All three sample size formulae were verified by simulation studies as described in [5-7]. Although these simulations were conducted with unknown variance components, the simulation-based empirical statistical power estimates were very close to those computed from the formulae presented in this chapter. Therefore, derivation of power function with unknown variances may not be necessary even for small N_3, although it might be possible through application of CDFs of central and non-central t distributions [22] replacing the standard normal CDF Φ and its inverse Φ^{-1} in equation in sample formulas for $N_3(\Delta_{(I)})$ (4..8), $N_3(\Delta_{(II)})$ (5.7) and $N_3(\Delta_{(III)})$ (6.7). Regardless, sample variance components can be used in lieu of population variance components for the purposes of statistical power calculations and sample size estimation.

The effects of the level one and level two correlations, ρ_1 and ρ_2, on the sample size requirements depend on the hypothesis of interest. The sample size

$N_3(\Delta_{(I)})$ for testing Hypothesis (I) is an increasing function of both ρ_1 and ρ_2 since the variance inflation factor is an increasing function of both. In contrast, the sample size $N_3(\Delta_{(II)})$ for testing Hypothesis (II) is a *decreasing function* of ρ_1 and independent from ρ_2 since the variance of the slopes is a decreasing function of only ρ_1. The effects of ρ_1 and ρ_2 on the sample size $N_3(\Delta_{(III)})$ for testing Hypothesis (III) are somewhat complicated since the variance of the group mean difference at the end of a study under model (6.1) is a combination of the variance of the sample overall means and that of the estimated slope as shown in equation (6.4). The sample size $N_3(\Delta_{(III)})$ is an increasing function of ρ_2, which inflates the variance of the sample overall means but does not have any effect on that of the estimated slope. However, with respect to ρ_1, the former variance increases with increasing ρ_1 whereas the latter decreases with increasing ρ_1. Nevertheless, it can be seen in Table 3 that the required sample sizes $N_3(\Delta_{(III)})$ increase with increasing ρ_1 because the positive effect of ρ_1 on the $Var(\hat{\theta})$ is greater than its negative effect on $Var(\hat{\gamma})$ in equation (6.4).

When designing a cluster-RCT, several factors should be considered since the required samples sizes for the different levels depend on many design parameters. First, the choice of hypothesis must be made on the basis of the scientific questions and relevance to the field. Sample size should then be determined based on that specific hypothesis and the hypothesized effect. Still, determination of the optimal combination of the three sample sizes N_3, N_2, and N_1 should also be guided by the expected cost per unit for each level, the feasibility of recruitment and clinical assessments, as well as availability of resources. The sample size of a specific level may also be dictated by the nature of the disease under study. For example, in psychopharmacological trials, weekly assessments over a 2 month or longer follow-up period are not uncommon [23]. As a result, the number of repeated measurements on each subject (N_1) would necessarily have to be large even though power may not be optimal [24, 25]. In general, however, the number of higher level data units N_3 has a greater effect on power than that of lower level data units. This is apparent especially when comparing group means by noting that N_3 does not contribute to the variance inflation factor f (4.6) as shown here and elsewhere [26].

The sample sizes were derived under the assumption that there will be no missing units for all levels. However, since attrition of subjects during a trial in reality is the norm rather than the exception [27], the number of level two units will likely vary (i.e., $j = 1, 2, ..., n_i$, depending the i-th level three unit) as well as the number of level one units (i.e., $k = 1, 2, \ldots, n_{ij}$, depending on both level three and two units). If such variation is *completely at random* as in the missing data setting [28], one could replace the varying cluster sizes with the average cluster size, i.e., replacement of N_2 and N_1 by $\tilde{N}_2 = \sum_{i=1}^{N_3} n_i / N_3$ and $\tilde{N}_1 = \sum_{i=1}^{N_3}\sum_{j=1}^{n_i} n_{ij} / (\tilde{N}_2 N_3)$, respectively. This strategy was shown to be effective in an application of a two-level mixed effects logistic regression model to the analysis of a subject-based randomized clinical trial with a repeatedly measured binary outcome [29]. Nevertheless, Roy et al. [30] derived a general form for sample size determination using a mixed-effects linear model for three-level data, taking into account potential attrition rates and more general correlation structures.

In conclusion, presented in this chapter were closed-form formulae for determining sample sizes for designing 3-level cluster-randomized clinical trials that aim to test the significance of an overall intervention effect, time-by-intervention effect, and local intervention effect at the end of study. The proposed approaches can also be applied to two-level hierarchical data by simply setting $N_1 = 1$ in equations (4.8), (5.7) and (6.7), respectively. This reduces the three-level data structure to a two-level structure. Similarly, by setting $N_2 = N_1 = 1$, the power functions and sample sizes reduce to those for single-level data. Therefore, the methods in this chapter can easily accommodate lower level data structures as special cases. In addition, the general principles for determining the sample sizes of three-level cluster randomized trials can also be readily extended to design higher level cluster RCTs, such as four- or five-level trials.

References

[1] Hayes, RJ; Moulton, LH. *Cluster Randomized Trials*. CRC Press: Boca Raton, 2009.

[2] Donner, A; Klar, N. *Design and Analysis of Cluster Randomization Trials in Health Research*. Arnold: London, 2000.

[3] Donner, A; Klar, N. Statistical considerations in the design and analysis of community intervention trials. *Journal of Clinical Epidemiology* 1996, 49, 435-439.

[4] Donner, A; Klar, N. Pitfalls of and controversies in cluster randomization trials. *American Journal of Public Health,* 2004, 94, 416-422.

[5] Heo, M; Leon, AC. Statistical Power and Sample Size Requirements for Three Level Hierarchical Cluster Randomized Trials. *Biometrics,* 2008, 64, 1256-1262.

[6] Heo, M; Leon, AC. Sample size requirements to detect an intervention by

time interaction in longitudinal cluster randomized clinical trials. *Statistics in Medicine,* 2009, 28, 1017-1027.

[7] Heo, M; Kim, Y; Xue, XN; Kim, MY. Sample size requirement to detect an intervention effect at the end of follow-up in a longitudinal cluster randomized trial. *Statistics in Medicine,* 2010, 29, 382-390.

[8] Goldstein, H. *Multilevel Statistical Models.* (2nd edn). Wiley & Sons: New York, 1996.

[9] Raudenbush, SW; Bryk, AS. *Hierarchical Linear Models: Application and Data Analysis Methods.* (2nd edn). SAGE: Thousand Oaks, 2006.

[10] Hedeker, D; Gibbons, RD. *Longitudinal Data Analysis.* Wiley: Hoboken, NJ, 2006.

[11] Snijders, TAB; Bosker, RJ. *Multilevel Analysis: An Introduction to Basic and Advanced Multilevel Modeling.* SAGE: London, 1999.

[12] Serfling, RJ. *Approximation Theorems of Mathematical Statistics.* Wiley & Sons: New York, 1980.

[13] Cohen, J. The statistical power of abnormal-social psychological research: A review. *Journal of Abnormal and Social Psychology,* 1962, 65, 145-153.

[14] Cohen, J. *Statistical Power Analysis for the Behavioral Science.* Lawrence Erlbaum Associates: Hillsdale, NJ, 1988.

[15] Rosner, B. *Fundamentals of Biostatistics.* (6th edn). Duxbury: USA, 2006.

[16] Liu, AY; Shih, WJ; Gehan, E. Sample size and power determination for clustered repeated measurements. *Statistics in Medicine,* 2002, 21, 1787-1801.

[17] Diggle, PJ; Heagerty, P; Linag, KY; Zeger, SL. *Analysis of Longitudinal data.* (2nd edn). Oxford University Press: New York, 2002.

[18] Hsieh, FY. Sample size formulas for intervention studies with the cluster as unit of randomization. *Statistics in Medicine,* 1988, 7, 1195-1201.

[19] Bruce, ML; Ten Have, TR; Reynolds, CF; Katz, II; Schulberg, HC; Mulsant, BH; Brown, GK; McAvay, GJ; Pearson, JL; Alexopoulos, GS. Reducing suicidal ideation and depressive symptoms in depressed older primary care patients - A randomized controlled trial. *Jama-Journal of the American Medical Association,* 2004, 291, 1081-1091.

[20] Alexopoulos, GS; Katz, IR; Bruce, ML; Heo, M; Ten Have, T; Raue, P; Bogner, HR; Schulberg, HC; Mulsant, BH; Reynolds, CF; Grp, P. Remission in depressed geriatric primary care patients: A report from the PROSPECT study. *American Journal of Psychiatry,* 2005, 162, 718-724.

[21] Dietrich, AJ; Oxman, TE; Williams, JW; Schulberg, HC; Bruce, ML; Lee, PW; Barry, S; Raue, PJ; Lefever, JJ; Heo, M; Rost, K; Kroenke, K; Gerrity, M; Nutting, PA. Re-engineering systems for the treatment of depression in

primary care: cluster randomised controlled trial. *British Medical Journal,* 2004, 329, 602-605.

[22] Johnson, NL; Kotz, S. *Distributions in Statistics: Continuous Univariate Distributions-2.* Houghton Mifflin: New York, 1970.

[23] Heo, M; Papademetriou, E; Meyers, BS. Design characteristics that influence attrition in geriatric antidepressant trials: meta-analysis. *International Journal of Geriatric Psychiatry,* 2009, 24, 990-1001.

[24] Tan, FES; Berger, MPF. Optimal allocation of time points for the random effects model. *Communications in Statistics-Simulation and Computation,* 1999, 28, 517-540.

[25] Winkens, B; Schouten, HJA; van Breukelen, GJP; Berger, MPF. Optimal designs for clinical trials with second-order polynomial treatment effects. *Statistical Methods in Medical Research,* 2007, 16, 523-537.

[26] Teerenstra, S; Moerbeek, M; van Achterberg, T; Pelzer, BJ; Borm, GF. Sample size calculations for 3-level cluster randomized trials. *Clinical Trials,* 2008, 5, 486-495.

[27] Heo, M; Leon, AC; Meyers, B; Alexopoulos, GS. Problems in statistical analysis of attrition in randomized controlled clinical trials of antidepressants for geriatric depression. *Current Psychiatry Reviews* 2007, 3, 178-185.

[28] Little, RJA; Rubin, DB. *Statistical Analysis with Missing Data.* (2nd edn). Wiley: Hoboken, NJ, 2002.

[29] Heo, M; Leon, AC. Performance of a mixed effects logistic regression model for binary outcomes with unequal cluster size. *Journal of Biopharmaceutical Statistics,* 2005, 15, 513-526.

[30] Roy, A; Bhaumik, DK; Aryal, S; Gibbons, RD. Sample size determination for hierarchical longitudinal designs with differential attrition rates. *Biometrics* 2007, 63, 699-707.

In: Biometrics: Theory, Applications and Issues ISBN: 978-1-61728-765-7
Editor: Ellen R. Nichols, pp. 29-56

Chapter 2

A MODIFIED LOCALLY LINEAR DISCRIMINANT EMBEDDING FOR TUMOR CLASSIFICATION

Shanwen Zhang, Deshuang Huang and Bo Li
Hefei Institute of Intelligent Machines, Chinese
Academy of Sciences, Hefei, Anhui, China

Abstract

One important application of gene expression profiles data is tumor classification. Because of its characteristics of high dimensionality and small sample size problem, and a great number of redundant genes not related to tumor phenotypes, various dimensional reduction methods are currently used for gene expression profiles preprocessing. Manifold learning is a recently developed technique for dimensional reduction, which are likely to be more suitable for gene expression profiles analysis. This chapter will focus on using manifold learning methods to nonlinearly map the gene expression data to low dimensions and reveal the intrinsic distribution of the original data, so as to classify the tumor genes more accurately. Based on Locally Linear Embedding (LLE) and modified maximizing margin criterion (MMMC), a modified locally linear discriminant embedding (MLLDE) is proposed for tumor classification. In the proposed algorithm, we design a novel geodesic distance measure, and construct a vector translation and distance rescaling model to enhance the recognition ability of the original LLE from two aspects. One is the property that the embedding cost function is invariant to translation and rescaling, the other is that the transformation to maximize MMMC is introduced. To validate the efficiency, the proposed algorithm is applied to classifying seven public gene expression datasets. The experimental results show that MLLDE can obtain higher classification accuracies than some other methods.

Keywords: Locally linear embedding (LLE); Weighted local linear smoothing (WLLS); Modified maximizing margin criterion (MMMC); Modified locally linear discriminant embedding (MLLDE).

1. Introduction

Although tumor classification is by no means a new subject in the application domain of microarray technique, highly accurate tumor classification is difficult to achieve. The main reason is that the number n of samples collected is relatively small compared to the number p of genes per sample which are usually in the thousands. In statistical terms the very large number of variables (genes) compared to a small number of samples (microarrays) make most of classical 'class prediction' methods difficult to employ. And among the large number of genes, only a small part may benefit the correct classification of tumor subtypes. That is, most of genes have little or no contribution for the classification. Even worse, some genes may act as "noise" and depress the classification accuracy. These characteristics usually result in the known problems of "curse of dimensionality" and over-fitting of the training data for traditional classification methods. An efficient way to solve this problem is by using dimension reduction techniques in conjunction with discriminant procedures.

Dimensional reduction is an important preprocessing step before classifying multidimensional data. For dimensional reduction, there are mainly two different research interests, i.e., linear dimensionality reduction (LDR) methods and nonlinear dimensionality reduction (NLDR) methods. Although some of LDR methods have been used efficiently for tumor classification [1-6], classical LDR methods can only find flat Euclidean structures, but fail to discover the curved and nonlinear structures of the input data. Nonlinear mapping techniques, such as self-organizing maps (SOM) [7] and topology preserving networks [8], do not involve a global cost function, tend to have many free parameters, and are limited in low-dimensional data sets [9]. To overcome these drawbacks, several manifold learning based algorithms, e.g., Isometric Feature Mapping (Isomap) [10] and Locally Linear Embedding (LLE) [11], have been developed. These methods aim at finding the meaningful nonlinear low dimensional structures hidden in the high dimensional data. The basic assumption of these algorithms is that the input data lie on or close to a low-dimensional nonlinear manifold.

LLE is a representative, popular, unsupervised local linear manifold learning approach. The main principle of the LLE algorithm is to preserve the local geometry of high dimensional data points in both the embedding space and the intrinsic space. Each sample in the observation space is a linearly weighted average of its neighbors. It is based on the simple geometric intuitions: (1) each high-dimensional data point and its neighbors lie on or close to a locally linear patch of a manifold, and (2) the local geometric characterization in original data space is unchanged in the output data space. Although LLE is a novel dimensional reduction approach, a lot of limitations are exposed when it is applied to classification. One is that LLE is sensitive to outliers and noise. Due to the locality geometry preservation, LLE is not robust against outliers in the data and are in general sensitive to noise. Often they fail in the presence of the high dimensional outliers or noise, because the outliers or noise may change the local structure of the manifold. When the local outliers or noise level are increased, mappings quickly become very poor. Another is the *out-of-sample problem*. Because the weighted matrix of LLE is constructed on the training data, when dealing with a new data point, how to generalize the results of the training samples to the new data is a problem need to be solved. Third is that the classical LLE neglects the class information, which will impair the classification accuracy. More serious, LLE may lose its efficiency in dimensional reduction for classification since it is built based on Euclidean distance for exploiting neighborhood information, and since the local Euclidean distance does not match the classification property generally. That is to say, two sample points belonging to different classes may also have a short Euclidean distance. This phenomenon may result in that the neighbors of a point may come from different classes. In order to remedy the shortages of LLE, many modified LLE algorithms have been boomed by using the sample label information, some of them have been used for tumor classification [12-14].

Hadid et al. [15] proposed a robust LLE. This method is based on the assumption that all outliers are very far away from the data on the manifold. But this assumption is not always true for many real-world applications. Based on a weighted version of PCA, Zhang et al. [16] developed weighted local linear smoothing (WLLS) for outlier removal and noise reduction for a set of noisy points sampled from a nonlinear manifold. This method can be used by LLE as a preprocessing procedure so as to obtain a more accurate reconstruction of the underlying nonlinear manifolds. However, the method for determining the weights is heuristic in nature without formal justification. Chang et al. [17] proposed a robust LLE based on the robust PCA, which is robust in the presence of the large noises. This method would also fail when the sample is low or the data points are unevenly sampled. Hein et al. [18] proposed a denoising method based on graph Laplacian as preprocessing step for further manifold learning. This method has good performance for noisy data, but when the outliers are closer to another digit component, wrong transformation will happen. Zhang et al. [19] presented a multiple weights LLE. Although, it could solve the low density sample problem, it would fail in the case as the data points are not distributed on or close to a 2D or 3D nonlinear manifold. Yin et al. [20] proposed a neighbor smoothing embedding (NSE) for noisy data. It can efficiently maintain an accurate low-dimensional representation of the noisy data with less distortion, and give higher average classification rates compared to others. But this method introduces an additional parameter and ignores the statistical feature. Pan et al. [21] proposed a weighted locally linear embedding (WLLE) to discover the intrinsic structures of data. It has a good performance for both uniform and non-uniform distribution. But in WLLE, the weighted distance is not always the geodesic distance in real-data.

Zhang et al. [22] presented a unified framework of LLE and LDA, which can be used for tumor classification. This framework essentially equals to LLE+LDA. There are still some weaknesses in this proposed algorithm. In the whole process, some useful embedding information may be probably thrown away, since the dimensionality is reduced twice. Kim et al. [23] presented a locally linear discriminant analysis (LLDA) which involves a set of locally linear transformations. In this method, the global nonlinear data structures are locally linear and local structures can be linearly aligned, input vectors are projected into each local feature space by linear transformations, which maximize the between-class covariance while minimizing the within-class covariance. Ridder et al. [24] introduced a supervised LLE (SLLE) to deal with data sets containing multiple (often disjoint) manifolds, where the Euclidean distance is simply enlarged by adding a constant for the pairs of points belonging different classes, keeping others unchanged. However, this strategy cannot be used for testing data. As a replacing compensation, Euclidean distance has to be used for selecting neighbors of testing data. The incoherency of neighborhood selecting method for training data and testing data may affect the classification results negatively. Zhao et al. [13] presented a novel dimensional reduction method for classification based on probability LLE (PLLE). In this method, logistic discrimination (LD) is adopted for estimating the probability distribution as well as for classification on the reduced data. Different from the SLLE that is only used for the dimensional reduction of training data, PLLE can be applied on both training and testing data. Applying LD to construct the probability function is a key step of PLLE. But the LD itself is a more complicated problem.

In gene expression profiles analysis, the dimensional reduction for tumor classification aims to project the gene expression data into a feature space in which the samples from different classes could be clearly separated. To achieve this aim, in this chapter we propose a supervised LLE algorithm, named as modified locally linear discriminant embedding (MLLDE). The goal of MLLDE is to take full advantage of the tumor class information to improve the tumor classification ability. Owing to modified Maximum Margin Criterion (MMMC) and the geodesic distance strategy, the contributions of this chapter are stated as follows:

a) Weighted local linear smoothing is applied as a preprocessing procedure for outlier removal and noise reduction.
b) MMMC is presented to overcome the *small size sample* problems.
c) Linear transformation $Y=A^T X$ is introduced to overcome the *out-of-sample* problems.
d) Geodesic distance metric is designed to enlarge the Euclidean distance of inter-class points.
e) MLLDE is proposed and is applied to tumor classification.

2. Locally Linear Embedding (LLE)

LLE constructs a neighborhood-preserving mapping based on the weight matrix. As a result, LLE can preserve the local manifold structure of one face class in the dimensionality reduced space. To establish the mapping relationship between the observed data and the corresponding low-dimensional data, the LLE algorithm is used to obtain the corresponding low-dimensional data Y of the original training set X.

Let $X=[X_1, X_2,..., X_n]\in R^{D\times n}$ be a set of n data points in a high dimensional data space R^D. The data points are assumed to lie on or near a nonlinear manifold of intrinsic dimensionality $d<D$ (generally $d\ll D$). Provided that sufficient data are available by sampling well from the manifold, the goal of LLE is to find a low-dimensional embedding of X by mapping the D-dimensional data into a single global coordinate system in R^d. Let us denote the corresponding set of n points in the embedding space R^d by $Y=[Y_1, Y_2,..., Y_n]\in R^{d\times n}$. The key steps of LLE algorithm are described as follows:

Step 1: For each data point $X_i\in X$, find the set Δ_i of k nearest neighbors of X_i by K-NN algorithm or $\varepsilon-ball$ algorithm in Euclidean space.

Step 2: Compute the reconstruction weights W_{ij} of the neighbors that minimize the error ε_i of reconstructing X_i, i.e., seek the best reconstruction weights.

$$\varepsilon_i(W) = \arg\min \left\| X_i - \sum_{X_j \in \Delta_i} W_{ij} X_j \right\|^2 \tag{1}$$

where $\|\bullet\|$ is the Euclidean norm.

Intuitively, W_{ij} reveals the layout of the points around X_i, it subject to two constraints, $\sum_{X_j \in X_i} W_{ij} = 1$ and $W_{ij}=0$ for any $X_j \notin \Delta_i$. In the first constraint, each point is represented as a convex combination of its neighbor; the second reflects that LLE is a *local* method.

Clearly, minimizing $\varepsilon_i(W)$ subject to the above constraints is a constrained least squares problem. After repeating *Step* 1 and two constraints to all the n data points, the reconstruction weights consist of a weight matrix $W=[W_{ij}]_{n\times n}$.

Step 3: Compute the low-dimensional embedding Y for X that best preserves the local geometry represented by the reconstruction weights. This means to solve Eq. (2),

$$\varepsilon(Y) = \arg\min \sum_{i=1}^{n} \left\| Y_i - \sum_{X_j \in \Delta_i} W_{ij} Y_j \right\|^2 \tag{2}$$

where Y_i subject to two constraints, i.e. $\sum_{i=1}^{n} Y_i = 0$ (translation invariant embedding) and $\frac{1}{n}\sum_{i=1}^{n} Y_i Y_i^T = I$ (normalized unit covariance), 0 is a column vector of zeros and I is an identity matrix. Two constraints are appended to remove the translational degree of freedom and the rotational degree of freedom, respectively.

Based on the matrix W, we can define a sparse, symmetric, and positive semi-definite matrix M based on $M=(I-W)^T(I-W)$, then *Step* 3 is transformed to the solution of eigenvector decomposition which can be seen as follows:

$$\begin{aligned}\varepsilon(Y) &= \arg\min \sum_i \left\| Y_i - \sum_{X_j \in \Delta_i} W_{ij} Y_j \right\|^2 \\ &= \arg\min \left\| (I - W)Y \right\|^2 \\ &= \arg\min [(I - W)Y]^T (I - W)Y \\ &= \arg\min Y^T (I - W)^T (I - W)Y \\ &= \min trace(YMY^T) \end{aligned} \quad (3)$$

In Eq. (3), *trace* denotes the trace of a square matrix, i.e. the sum of the diagonal elements. By the Rayleigh-Ritz theorem [25], minimizing Eq. (3) can be performed by finding the eigenvectors with the smallest (nonzero) eigenvalues of the sparse matrix *M*. This corresponds to solving a sparse eigenvalue problem, using singular value decomposition (SVD) without performing a full diagonally matrix, to find the d+1 eigenvectors of M with the smallest $d+1$ eigenvalues.

3. Noise Influence upon Manifold Learning

As to noise influence upon manifold learning, it is already described in some correlated literatures [16-18]. The following are two actual examples indicating that noise affects manifold dimensionality reduction. These data belong to artificial data. Figure 1 shows how the classical LLE works in finding the low dimensional embedding of the S-curve and Swiss-roll manifold from R^3 to R^2. Figure 1 (A) are original data of S-curve and Swiss-roll; Figure 1 (B) are S-curve and Swiss-roll data to which Gaussian noise is added; Figure 1 (C) are the LLE embedding results of the noisy data in (B).

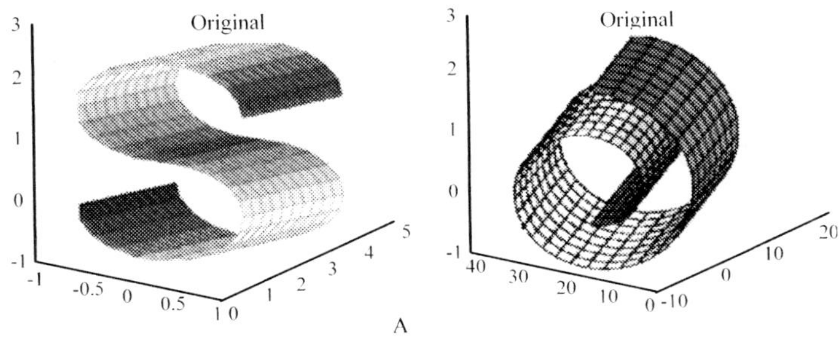

Figure 1. Continued on next page.

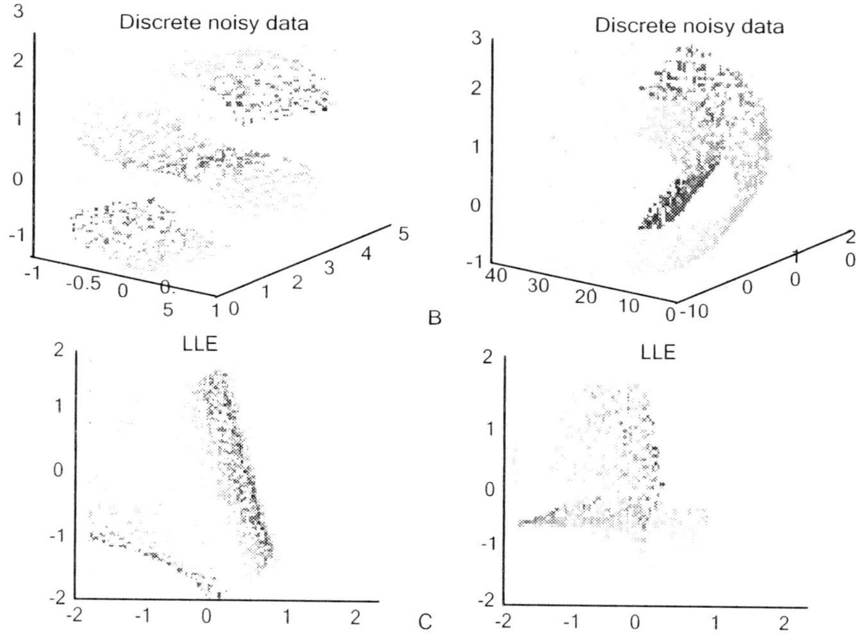

Figure 1. LLE applied to the S-curve and Swiss-roll dataset, respectively. (A) S-curve and Swiss-roll manifold in R^3 ; (B) Discrete points randomly sampled from the manifold (totally 1,500 clean points from S-curve and 150 noisy points added; totally 1,800 clean points from Swiss-roll and 200 noisy points added); (C) LLE result embedding space in R^2 with the nearest neighbors k=15 based on Euclidean distance.

As we can see clearly from Figure 1(C), the LLE algorithm is topologically unstable; it cannot preserve well the local geometry of the data manifolds in the embedding space in the presence of noise, the entire data topological structure is also destroyed. By further experiments, we are able to find that small errors in the data connectivity (topology) can lead to large errors in the solution. Choosing a very small neighborhood is not a satisfactory solution, as this can detach the manifold into a large number of disconnected regions. In the presence of outliers (noises), the K-nearest neighbors of a neat point on the manifold may no longer lie on a locally linear patch of the manifold, leading to a small bias to the reconstruction. As for an outlier (noise) point, its neighborhood is typically much larger than that of a clean data point.

Through observing the embedding results of two above noisy data, we can see that noise influence upon manifold data mainly reflects the destruction of topological structure; such destruction means the failure of manifold algorithm, therefore the key problem of whether manifold learning can be used successfully is how to validly deal with noise in initial data.

4. Modified Locally Linear Discriminant Embedding

In this section, the proposed modified locally linear discriminant embedding (MLLDE) algorithm is presented by combining LLE with modified maximize margin criterion (MMMC).

4.1. Geodesic Distance Definition

We represent the microarray data as points in a high-dimensional space. Points, either genes or samples, which are close to each other in this space which is considered more similar than points far from each other. An important step in the mathematical modeling is how to define the similarity between the points. The ordinary Euclidean distance measure is not necessarily the most suitable similarity measure in the microarray data space. A good distance measure should reflect the biological reality, so that two genes, or two samples, close to each other should also be similar in some biological sense. Different distance measures might be suitable for different applications. A difficulty here is that it is not always clear how to define biological similarity. So far a variety of distance measures have been used in the literature. For data lying on a nonlinear manifold, the "true distance" between two data points is the geodesic distance on the manifold, i.e. the distance along the surface of the manifold, rather than the straight-line Euclidean distance. The approximation of geodesic distance includes two cases. For the neighboring points, Euclidean distance in the input space provides a good approximation to geodesic distance. For faraway points, geodesic distance can be approximated by adding up a sequence of "short hops" between neighboring points.

For a pair of different points X_i and X_j, we define their class-similarity by

$$\delta(X_i, X_j) = \begin{cases} 1, & \text{if } C_i = C_j \\ 0, & \text{if } C_i \neq C_j \end{cases} \tag{4}$$

where C_i is the class label of X_i .

The geodesic distance metric introduced in MLLDE algorithm is defined as

$$L(X_i, X_j) = \|X_i - X_j\| + \beta \cdot l \cdot (1 - \delta(X_i, X_j)) \tag{5}$$

where $l = \max_{ij} \|X_i - X_j\|$ is the data diameter in Euclidean distance, $\beta \in [0,1]$ is a tuning parameter that controls the amount to which class information should be incorporated.

From Eq. (5), with the class information, the properties of the $L(X_i,X_j)$ can be summarized as follows:

(1) With the increasing of the Euclidean distance, the intra-class $L(X_i,X_j)$ is equal to the Euclidean distance, while the inter-class $L(X_i,X_j)$ increases faster;
(2) When the Euclidean distance is equal, the inter-class $L(X_i,X_j)$ is larger than the intra-class $L(X_i,X_j)$, which reflects the class information of data points. That is to say, two points with smaller distance most likely have the same label. On the contrary, the distance between two points with different labels will be larger. This character is in favor of sample classification.

4.2. Modified Maximize Margin Criterion for Classification

The proposed MLLDE needs a criterion used for automatically finding an optimal linear transformation for classification. Maximum Margin Criterion (MMC) based dimensional reduction method is more efficient than LDA for calculating the discriminant vectors since it does not need to calculate the inverse within-class scatter matrix, which can successfully overcome the *small sample size problem*.

When performing dimensionality reduction, we want to find a (linear or nonlinear) mapping from the measurement space to some feature space such that the objective function J is maximized after the transformation. Consider a linear mapping A. We would like to maximize

$$J_1 = \max trace\{(S_b - S_w)\} \tag{6}$$

where $S_b = \sum_{i=1}^{c} p_i (m_i - m)(m_i - m)^T$ and $S_w = \sum_{i=1}^{c} p_i S_i$ are the between-class scatter matrix and the within-class scatter matrix in feature space, respectively; S_i is the intra-class scatter matrix of class i ; c is the number of classes; m_i and p_i are the mean vector and a priori probability of class i, respectively; m is the mean or centroid of all samples, $m_i = \sum_{j=1}^{n} x_{ij} / n$, $m = \sum_{i=1}^{c} p_i m_i$, $A = (A_1, A_2, ..., A_n)$ is projection matrix under the constraint $A_j^T A_j = 1$, $j = 1, 2, ..., n$.

Since A is a linear mapping, it is easy to obtain $S_b = A^T S_b A$, $S_w = A^T S_w A$. Then, we have

$$J_2(W) = \max trace\left\{A^T (S_b - S_w) A\right\} \tag{7}$$

The advantage of using the MMC is that it is unnecessary to compute the inverse of S_w, hence the singularity problem can be easily avoided. But its process is not always optimal for the classification instance. In order to take advantage of the property of the weighted matrix's invariance to the rescaling of LLE, we introduce a rescaling parameter α >0, and first translate the data to suitable places, then rescale the data with the same label to their centroids and all the centroids are kept unchanged [26,27]. So S_b is still preserved while S_w is rescaled, i.e. $S_b' = S_b$, $S_w' = \alpha S_w$.

Because the MMC only considers the global Euclidean structure of the data space and ignores any nonlinear manifold structure, a modified MMC (MMMC) can be proposed and rewritten as:

$$J_3(A) = \max trace\{A^T (S_b - \alpha S_w) A\} \tag{8}$$

By changing α, one can control the importance of the variance of the between-class data to that of the within-class data distributions. The function J_3 can efficiently be optimized once α is selected.

Since $trace(S_b)$ measures the overall variance of the class mean vectors, a large $trace(S_b)$ implies that the class mean vectors scatter in a large space. On the contrary, a small $trace(S_w)$ implies that every class has a small spread. Thus, a large $J_1(A)$, $J_2(A)$ or $J_3(A)$ indicates that patterns are close to each other if they are from the same class but are far from each other if they are from different classes. Under such instance, the distance between different centroids will be larger and the within class scatters will be smaller. Since the objective functions, $J_1(A)$, $J_2(A)$ and $J_3(A)$, maximize the between-class scatter S_b while minimizing the within-class scatter S_w in the locally transformed feature space, finding an optimal linear subspace for classifying means to maximize the function $J_1(A)$, $J_2(A)$ or $J_3(A)$.

4.3. MLLDE Algorithm

Based on the analysis mentioned above, it can be found that the linear approximation to the original LLE explores a linear subspace with the least reconstructed error. The linear approximation can improve the discriminability of the data. At the same time, the MMMC presented above can map the data into an optimal subspace for classification. That is to say, if the linear transformation obtained by linearized LLE can satisfy Eq. (7) or Eq. (8), the discriminability of the data will be improved greatly.

In MLLDE, k nearest-neighbors of each training point X_i are determined using the distance measure defined by Eq.(5) at first. In order to overcome the *out-of-sample* problem, in the linear version of LLE, a linear transformation $Y=A^T X$ is introduced, which aims to project a pattern closer to those with the same class but farther from patterns in different labels, thus the discriminant component after performing MLLDE can be represented as $[Y\text{-}A]$, which can be also represented by a linear transformation, i.e. $Y\text{-}T= A^TX$, where T denotes the translation matrix. Then the objective function of the original LLE is changed into $J_4(A)=\min trace\left\{YMY^T\right\}=\min trace\left\{A^T XMX^T A\right\}$ with two constrains $\sum_{i=1}^{n} A^T X=0$ and $\sum_{i=1}^{n} Y_i Y_i^T=A^T XX^T A=n\cdot I^T$. We discarded the first constrain because it will remove translation degree of freedom. At the same time, by the MMMC, the objective function is changed into $J_3(A)=\max trace\left\{A^T(S_b-\alpha S_w)A\right\}$. Thus we must solve a two-object optimal problem,

$$\begin{cases} J_4(A)=\min trace\left\{A^T XMX^T A\right\} \\ J_3(A)=\max trace\left\{A^T(S_b-\alpha S_w)A\right\} \\ subject\ to\ A^T XX^T A=n\cdot I \end{cases} \tag{9}$$

This problem can be easily changed into a general objective optimized problem,

$$\begin{cases} \min trace\left\{A^T\left(XMX^T-(S_b-\alpha S_w)\right)A\right\} \\ A^T XX^T A=n\cdot I \end{cases} \tag{10}$$

It can be solved by Lagrangian multiplier,

$$\frac{\partial}{\partial A} trace\left\{A^T\left(XMX^T-(S_b-\alpha S_w)\right)A-\lambda\left(A^T XX^T A-n\cdot I\right)\right\}=0 \tag{11}$$

Then we obtain an equation,

$$\left(XMX^T-(S_b-\alpha S_w)\right)A=\lambda XX^T A \tag{12}$$

where $\lambda=\{\lambda_1,\lambda_2,...\}, A=\{A_1,A_2,...\}$, λ_i is regarded as the generalized ith eigenvalue of $\left(XMX^T-(S_b-\alpha S_w)\right)$ and XX^T , A_i is the corresponding eigenvector. Therefore, the objective function is maximized when A is composed of the first d smallest eigenvectors of the above generalized eigenvalue decomposition.

From above analysis, the steps of MLLDE algorithm can be summarized as follows:

Step 1: Define a distance metric according to Eq. (5).

*Step*2: For each sample data X_i, compute a set $N(X_i)$ of k nearest neighbors by *K-NN* algorithm or $\varepsilon-ball$ algorithm.

Step 3: Find the weight of linear coefficients. Weights are assigned to each edge so as to minimize the error,

$$\varepsilon_i\left(W\right)=\arg\min\left\|X_i-\sum_{X_j\in N(X_i)} W_{ij}X_j\right\|^2.$$

Step 4: Repeat *Step* 3, constitute a sparse matrix $W=[W_{ij}]_{n\times n}$ that encodes local geodesic properties specifying the relation of each X_i in terms of its nearest neighbors.

Step 5: Construct a matrix M based on $M=(I-W)^T(I-W)$ and construct a matrix XMX^T.

Step 6: Compute two scatter matrices S_b, S_w and their weighted difference $S_b-\alpha S_w$, respectively. Then construct a matrix, $U=\left(XMX^T-(S_b-\alpha S_w), XX^T\right)$.

Step 7: Compute eigenvalue decomposition for the matrix U, remove the zero eigenvalue, and compute the d bottom generalized eigenvalues and the corresponding eigenvectors matrix V, and then obtain d dimensional embedding $Y=V^T X$.

Step 8: Adopt optimal classifier to classify the embedding results.

4.4. MLLDE for Classification

Step 1: Remove outlier and reduce noise based on weighted local linear smoothing for a set of noisy points sampled from a nonlinear manifold [16].

Step 2: Compute the low-dimensional projection of training data using MLLDE.

Step 3: Output the final linear transformation matrix.

Step 4: Project the new testing data points into low-dimensional discriminating subspace.

Step 5: Predict the corresponding class labels using a suitable classifier.

5. Results and Discussion

Applied to microarray data, classification can be used to, given an expression pattern from a tumor sample, make a diagnosis about what kind of cancer the patient suffers from. In the case of genes, classification can be used to discover new members of functional groups. In this section, the classification performance of MLLDE is evaluated on seven different data sets and compared with the performances of locally linear discriminant analysis (LLDA) [14], supervised locally linear embedding (SLLE) [12, 24] and probability-based locally linear embedding (PLLE) [13].

5.1. Datasets

We applied our proposed method on seven publicly available data sets to evaluate its performance. They are Small Round Blue Cell (SRBC) tumor of Khan et al. [28], Acute Leukemia of Golub et al. [29], High-Grade Glioma (HGG) of Nutt et al. [30], Breast Tumor of Van't Veer et al. [31], Lung tumor of Gordon et al. [32], Colon of Alon et al. [33], and Central Nervous System (CNS) tumor of Pomeroy et al. [34]. The detailed descriptions of seven data sets are explained as follows.

1. SRBCT data set. From the web site: http://research.nhgri.nih.gov/microarray/Supplement, we download the SRBCT dataset which contains 88 samples with 2,308 genes in every sample. According to the original literature, there are 63 training samples and 25 testing samples which contain five non tumor-related samples. The 63 training samples

contain 23 Ewing family of tumors (EWS), 20 rhabdomyosarcoma (RMS), 12 neuroblastoma (NB), and eight Burkitt lymphomas (BL). The test samples contain six EWS, five RMS, six NB, three BL, and five non tumor-related samples. The five non tumor-related samples are removed in our experiments.

2. Acute Leukemia data set. The acute leukemia data set published in 1999. Each data point has 7129 feature genes. The data has 38 bone marrow samples from adult patients as training points of which 27 samples are acute lymphoblastic leukemia (ALL) and the other 11 samples are acute myeloid leukemia (AML). The independent testing set consists of 34 samples from adults and children with 20 ALL and 14 AML, of which 24 samples are bone marrow and the other 10 points are peripheral blood specimens. Only 4 AML of the testing set are from adult patients.
3. Glioma data: 50 high-grade glioma samples were carefully selected, 28 glioblastomas and 22 anaplastic oligodendrogliomas, a total of 21 classic tumors was selected, and the remaining 29 samples were considered non-classic tumor. The training set consists of 21 gliomas with classic histology of which 14 are glioblastomas and 7 anaplastic oligodendrogliomas. The test set consists of 29 gliomas with non-classic histology of which 14 are glioblastomas and 15 are anaplastic oligodendrogliomas. The number of gene expression levels is 12625.
4. Breast Cancer data set. This data set is from Van't Veer et al. [31]. Each sample has 24,481 genes. The train set contains 78 patient samples, 34 of which are from patients who had developed distance metastases within 5 years (relapse), and the rest 44 samples are from patients who remained healthy from the disease after their initial diagnosis for interval of at least 5 years (non-relapse). Correspondingly, there are 12 relapse and 7 non-relapse samples as the testing points.
5. Lung Cancer data set. The Lung cancer data consists of 181 tissue samples of which 31 points are malignant pleural mesothelioma (MPM) and other 150 points are adenocarcinoma (ADCA) of the lung. Each sample is described by 12,533 genes. The training set contains 32 tissue samples (16 MPM and 16 ADCA), while the rest 149 samples are used for testing (15 MPM and 134 ADCA). Note that in this data set, the number of training samples is much less than the number of the testing ones. There may be no enough classification information from the training data and may be difficult to classify the testing points.
6. Colon data set. This data set collects 62 colonic tissues from colon-cancer patients; 40 tumor biopsies are from tumors (labelled as "negative") and 22

normal biopsies (labeled as "positive") are from healthy parts of the colons of the same patients. Each sample is represented by 2000 genes (selected from 6500 original genes based on the confidence in the measured expression levels). We choose the first 40 points containing 23 tumor biopsies and 17 normal biopsies as the training set. The rest 22 points containing 13 tumor biopsies and 9 normal biopsies form the testing set.

7. Central Nervous System (CNS). Another dataset is central nervous system tumors [34]. This dataset is composed of four types of central nervous system embryonal tumors. The dataset used in our experiment contains 5597 genes in 34 samples representing four distinct morphologies: 10 classic medulloblasomas, 10 malignant gliomas, 10 rhabdoids and 4 normals. In this experiment, we randomly choose 5 medulloblasomas, 5 malignant gliomas, 5 rhabdoids and 3 normals as training set, and the rest samples as test data.

All the data sets are normalized so that they have zero means and standard deviations. Then we implement outlier removal and noise reduction based on weighted local linear smoothing for a set of original tumor sample points sampled from a nonlinear manifold. Because these data sets are of high dimensionality and small sample size, as shown in Table 1, it is necessary to reduce the data dimensionality before applying the classifier.

Table 1. Gene expression profiles data sets

No.	Gene data	Gene	Sample	Training	Testing
1	SRBC tumor	2,308	83	63	20
2	Leukemia tumor	7,129	72	38	34
3	HGG Tumor	12,625	50	21	29
4	Breast Tumor	24,481	97	78	19
5	Lung Tumor	12,533	181	32	149
6	Colon Tumor	2,000	62	40	22
7	CNS Tumor	5,597	34	18	16

Since all data samples in these seven datasets have already been assigned to a training set or testing set, we can build the classification models using the training samples and estimated the classification accuracy using the testing set. To obtain reliable experimental results showing comparability and repeatability for different numerical experiments, this study not only uses the original division of each dataset in training and testing set, but also reshuffles all datasets randomly. In other words, all numerical experiments are performed with 20 random splitting of the seven original datasets. And, they are also stratified, which means that each randomized training and testing set contains the same amount of samples of each class compared with the original training and testing set.

5.2. Experimental Results

After LLDA, SLLE, PLLE and MLLDE have been applied to extracting features, different pattern classifiers can be adopted for classification, including K-NN [35], Bayesian [36], support vector machine [37,38], etc. In this study, we apply the 1-nearest neighborhood classifier for its simplicity. Then the experiments can be carried out to test the effectiveness of the proposed algorithm.

To illustrate the behavior of our proposed MLLDE algorithm depending on the reduction dimensionality d, the neighborhood size k, the tuning parameter β, and the rescaling coefficient α if necessary, many experiments are repeated with variant combinations of (d,k,β,α) for each data set.

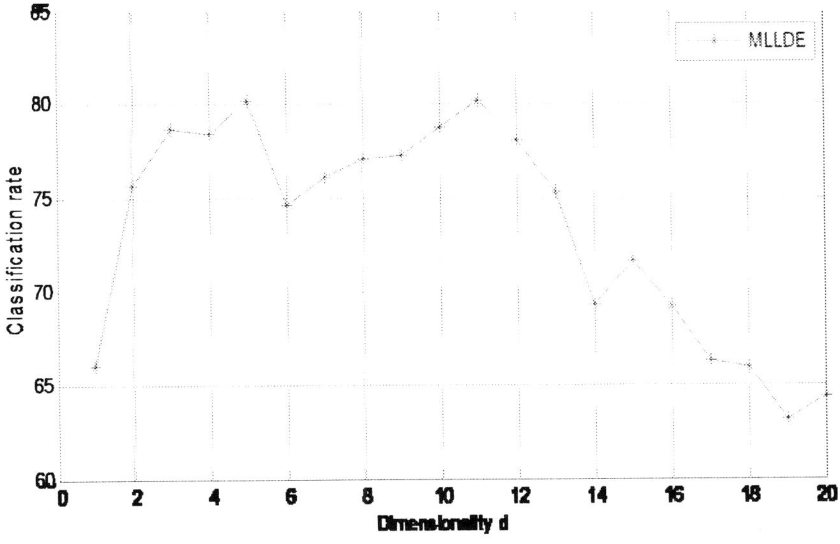

Figure 1. Classification rate with the varied d for High-Grade Glioma.

The parameters for most of the methods are determined empirically. That is, for each parameter, several values are tested and the best one is selected. In the phase of model selection, our goal is to determine proper parameters. Since it is very difficult to determine these parameters at the same time, a stepwise selection strategy is more feasible and thus is adopted here, i.e., one of three parameters is determined with respect to the other parameters being fixed or chosen.

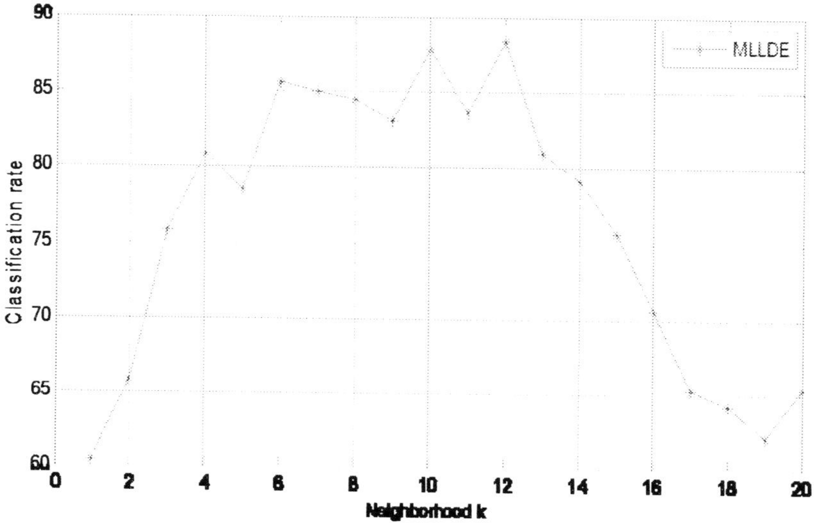

Figure 2. Classification rate with the varied k for High-Grade Glioma.

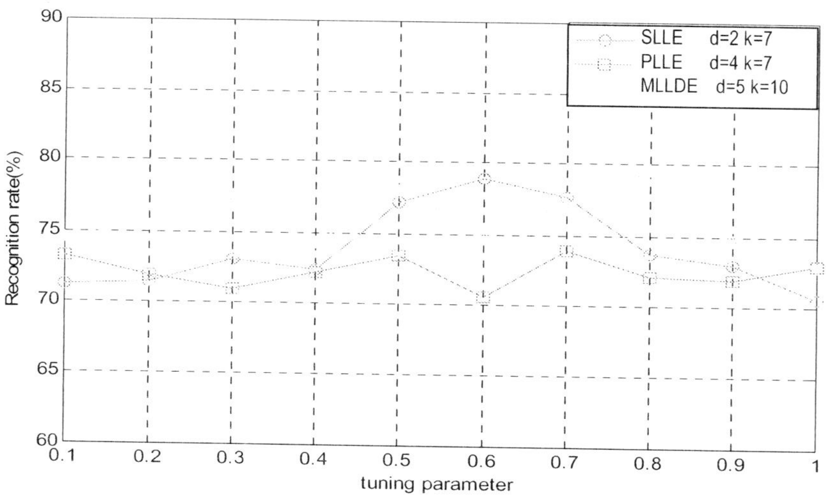

Figure 3. The best accuracy curves of three methods with respect to the tuning parameter β for High-Grade Glioma.

(1) The original divisions of High-Grade Glioma tumor and Lung tumor in training and testing set are used.

Firstly, we fix the parameters β=0.4, α=1 and k=4 for High-Grade Glioma data, then observe the classification effect of the proposed method versus the variation of the parameter d. The classification effect is shown in Figure 1.

Secondly, we fix the parameters β=0.4, α=1 and d=5 for High-Grade Glioma data, then observe the classification effect of proposed method versus the variation of the parameter k. The classification effect is shown in Figure 2.

Thirdly, to test the impact of tuning parameter β on the recognition rate, we plot the best accuracy curves of the High-Grade Glioma data via the tuning parameter β in Figure 3. The best result of SLLE is achieved at d=2, k=7; the best result of PLLE is achieved at d=4, k=7; while MLLDE has the best result at d=5; k=10.

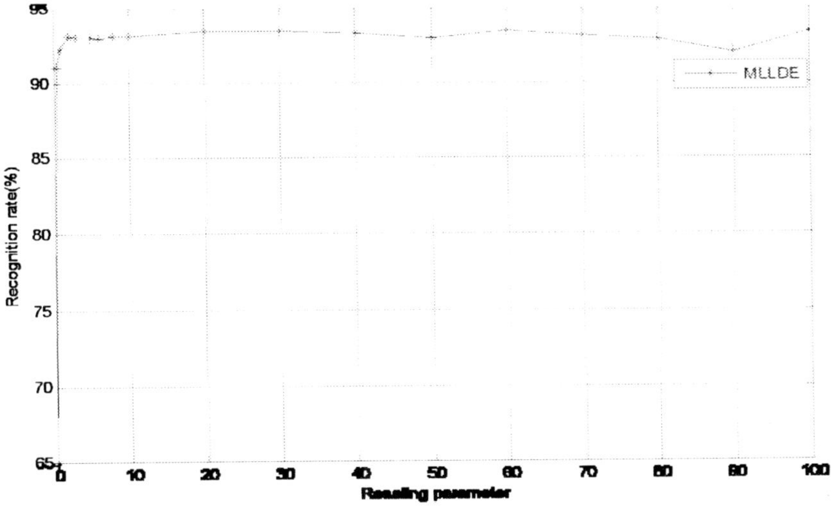

Figure 4. Classification rate with the varied rescaling coefficient α for Lung tumor data.

Table 2. Dimensions versus classification rate by varying rescaling coefficient α in MLLDE on Lung tumor data

Resaling Coefficient	0.01	0.1	1	20	100
Optimal accuracy rate	91.24	93.08	92.58	93.07	92.74
Dimension	16	14	17	16	15

At last, we fix the parameters β=0.8, d=5 and k=10 for Lung tumor data, then observe the classification effect of proposed method versus the variation of the rescaling parameter α. The classification effect is shown in Figure 4.

We further test the impact of rescaling coefficient α on the classification rate. The parameter α is set to 0.01, 0.1, 1, 20 and 100 for Lung tumor data, respectively. The optimal average classification rates for different coefficients are stated in Table 2, where the original division of Lung tumor dataset is used. The optimal classification rates can be obtained with different rescaling coefficients and the corresponding dimensions, for example, when α=0.1, the classification rate is 93.08% at 14 dimensions. However, the classification rate reaches 93.07% at 16 dimensions with α=20. The experimental results show that the parameter α has little affect on the classification rate, so in following experiments, α is fixed to 10 for the moment.

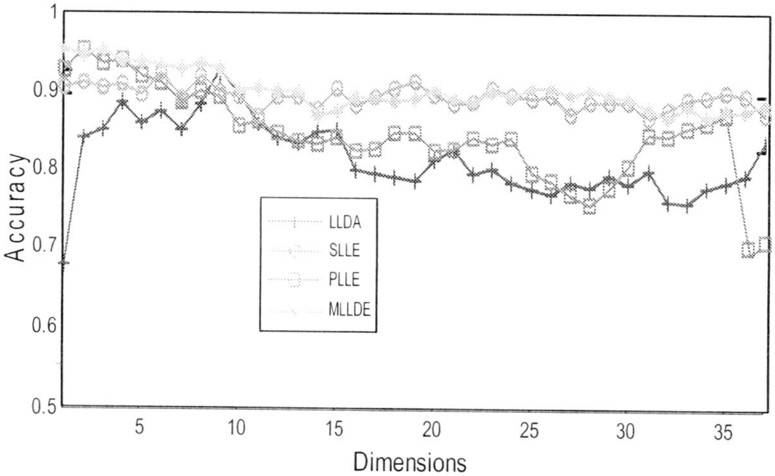

Figure 5. The mean accuracy on the test set of Leukemia data.

(2) The performances of four methods are illustrated on High-Grade Glioma tumor, Lung tumor and Central Nervous System tumor datasets.

To roundly show the experimental results versus the reduction dimensionality d, we show the classification accuracies and the corresponding standard deviations of four methods in Figure 5-Figure 10, where the dimensions are changed from 1 to the number of the training sample subtract from 1, the data are chosen from the original division and 20 random divisions, each accuracy rate and corresponding standard deviation are the mean of 21 experiments with a certain dimension. From Figure 5-Figure 10, we can see that the proposed method has best performance of tumor classification than three others. In addition, it is also found that there is no significant improvement if more dimensions are used.

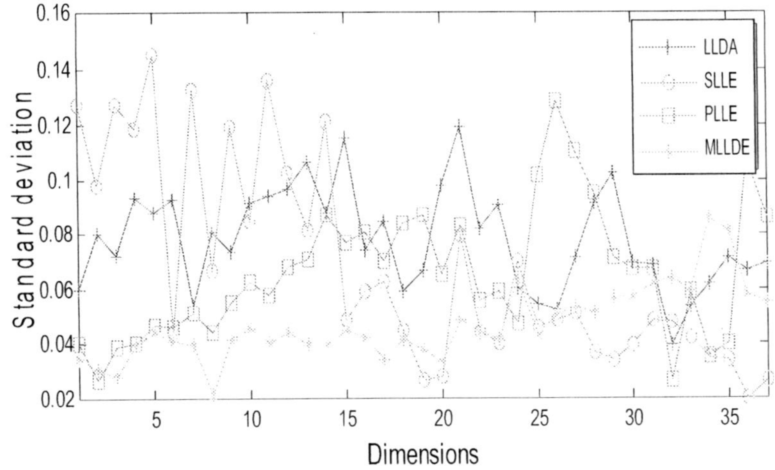

Figure 6. The corresponding standard deviation of Leukemia data.

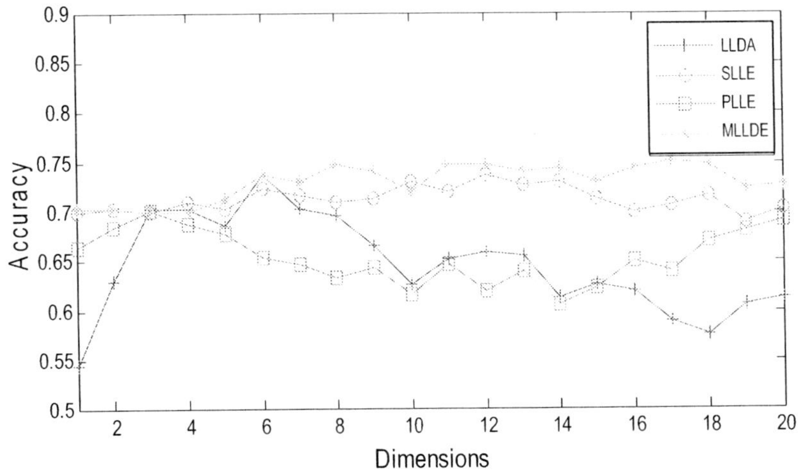

Figure 7. The mean accuracy on the test set of High-Grade Glioma.

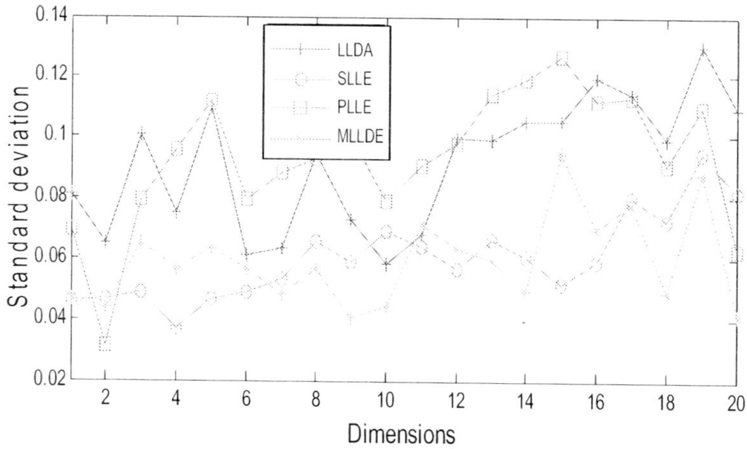

Figure 8. The corresponding standard deviation of High-Grade Glioma.

From above global-to-local search strategy, we can determine proper parameters for MLLDE. After globally searching over a wide range of the parameter space, we find a candidate interval where the optimal parameters might exist.

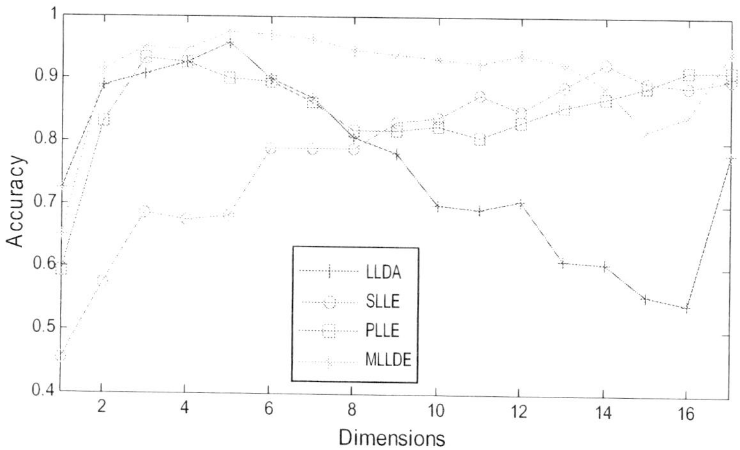

Figure 9. The mean accuracy on the test set of CNS tumor data.

In the following experiments, if not explicitly stated, we set k=2:2:20, d=1:10 for four methods, and set β=0.1:0.1:1 for SLLE, PLLE and MLLDE, and set the free parameter to 0.1:0.1:1 for LLDA, then there are total 10×10×10=1000 experiments for each of four methods. In each experiment, the data used are randomly chosen from the original division and 20 random divisions of each dataset in training set and testing set.

For each classification problem, the experimental results gave the statistical means and standard deviations, as shown in Table 3. Since the random splits for training set and testing set are disjoint, the results given in Table 3 should be unbiased. In Table 3, the classification results of testing data are only shown because the accuracies of training data here are almost 100% for all methods in almost all cases. It is found the MLLDE has higher accuracy and more stable than LLDA, SLLE and PLLE.

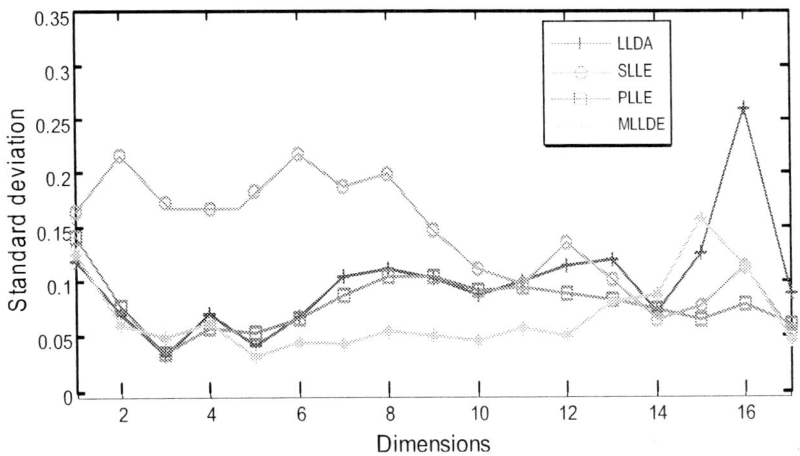

Figure 10. The corresponding standard deviation of CNS tumor data.

Table 3. The classification accuracy on testing set

Data	LLDA	SLLE	PLLE	MLLDE
SRBC	90.67±3.68	92.15±3.05	94.03±3.22	95.45±2.36
Leukemia	88.00±3.45	90.47±3.24	90.12±3.55	92.75±2.85
HGG	71.12±4.17	72.47±5.34	73.52±4.71	74.74±4.48
Breast	88.18±3.23	75.25±3.21	92.40±2.85	93.27±2.57
Lung	89.57±4.02	78.35±3.18	90.13±2.07	91.75±2.04
Colon	76.58±2.99	78.27±3.12	84.72±2.86	84.83±2.28
CNS	91.31±3.05	92.67±5.26	95.50±3.13	96.46±2.89

In above experiments, it is found that MLLDE is not only suitable for the gene expression profiles data with large training set and small testing set, but also suitable for the data with small training set and large testing set. It shows that MLLDE can better extract the key features of the original data sets in general different cases.

5.3. Discussion

From the experimental results mentioned above, we can find some interesting points as follows:

(1) The manifold learning methods are superior to some linear feature extraction methods. Moreover, compared to other supervised manifold learning techniques, MLLDE takes full advantages of the property of the original LLE, i.e., it is invariant to translations and rescaling, and the translation and rescaling can be automatically determined by an MMMC instead of being randomly set. The proposed MMMC aims to separate samples with different labels farther and cluster samples with the same label closer. Thus the proposed method can gain better recognition rate.

(2) Rescaling and translation are contained in the proposed algorithm. The between-scatter matrix S_b has been changed although the within-scatter matrix S_w has not changed after the linear translation. If we rescale the data set, the within-scatter matrix will also be changed. Compared to changing the within-scatter matrix, the contribution for improving the discriminability will be bigger by changing the between-scatter matrix. This is because rescaling cannot change the distances between centroids of different classes, i.e. S_b. In order to map the data belonging to different labels farther, a linear translation and a rescaling transform are taken into the proposed algorithm, which is a key to enhancing the discriminability of the data. Moreover, the rescaling is also adopted to cluster the data point closer, which helps the data points to be classified.

Of course, it's only the primary conclusion from our limited experiments. In the future more experiments should be done to verify the conclusion.

6. Conclusions

Dimensionality reduction is a mapping from high dimensional space to the one with fewer dimensions. Though many dimensional reduction methods have been proposed and widely applied in gene expression profiles analysis, there is no standard and uniform rule for gene expression profiles dimensional reduction because of its characteristics of small sample size and high dimensionality, and a great number of redundant genes as noise. There are many researchers who have been ceaselessly engaging in gene expression profiles analysis today. In this chapter, we aim to embed the gene expression profiles data points in as a low-dimensional space as possible in order to avoid the dimensionality curse and be useful for tumor classification. Based on the classical LLE, a supervised and discriminant method, namely MLLDE, is presented and applied to tumor classification. To validate the efficiency, the MLLDE was applied to tumor classification on seven different publicly available gene expression profiles datasets, with the 1-NN classifier and comparing with other three novels supervised dimensional reduction methods, LLDA, SLLE and PLLE. The proposed approach can effectively extract the most discriminant features. Compared to other feature extraction algorithms, the new technique does not suffer from the small sample size problem, the dimensionality reduction two times problem and the disconnected component problems. The experimental results show that the new method is effective.

Because the gene expression profiles data is very special, complicated, and nonlinear, the manifold learning may be a good choice for its analysis. Our work in this chapter may be a meaningful attempt in manifold and bioinformatics fields. Further research will address the problems of choosing suitable distance metric and parameters. Some aspects deserve further analysis, for example, the employment of different approximations for the mapping between the observed space and the embedding, in order to project new observations in the latent space, and how to define an optimal partition of the manifold into patches. So there should be much room for studying the application of manifold learning to bioinformatics.

References

[1] Antoniadis, A., Lambert-Lacroix, S. & Leblanc, F. (2003). Effective dimensional reduction methods for tumor classification using gene expression data. Bioinformatics, 19(0), 1-8.

[2] Gosh, D. (2002). Singular value decomposition regression modeling for classification of tumors from microarray experiments. In Proceedings of the Pacific Symposium on Biocomputing, 18-22.

[3] Huang, D. S. & Zheng, C. H. (2006). Independent component analysis-based penalized discriminant method for tumor classification using gene expression data. Bioinformatics, 22(15), 1855-1862.

[4] Nguyen, D. & Rocke, D. (2002). Tumor classification by partial least squares using microarray gene expression data. Bioinformatics, 18, 39-50.

[5] Teschendorff, A. E. (2005). et al: A variational Bayesian mixture modeling framework for cluster analysis of gene-expression data. Bioinformatics, 21, 3025-3033.

[6] Zheng, C. H. et al (2006). Nonnegative independent component analysis based on minimizing mutual information technique. Neurocomputing, 69, 878-883.

[7] Vidaurre, D. & Muruzabal, J. (2007). A quick assessment of topology preservation for SOM structures. IEEE Trans. Neural Netw, 18(5), 1524-1528.

[8] Martinets, T. & Schulten, K. (1994). Topology representing networks. Neural Netw, 7, 507-523.

[9] Lin, T., Zha, H. & Lee, S. U. (2006). Riemannian manifold learning for nonlinear dimensionality reduction. in Proc. 9th Eur. Conf. Comput. Vis, 1, 44-55.

[10] Tenenbaum, J. B., Silva, V. D. & Langford, J. C. (2000). A global geometric framework for nonlinear dimensionality reduction. Science 290 (5500), 2319-2323.

[11] Roweis, S. T. & Saul, L. K. (2000). Nonlinear dimensionality reduction by locally linear embedding. Science, 290, (5500), 2323-2326.

[12] Pillati, M. & Viroli, C. (2005). Supervised Locally Linear Embedding for Classification: An Application to Gene Expression Data Analysis. In: Proceedings of 29th Annual Conference of the of the German Classification Society, 15-18.

[13] Zhao, L. & Zhang, Z. (2008). Supervised locally linear embedding with probability-based distance for classification. Computers and Mathematics with Applications, *Available online*, 1-8.

[14] Zheng, C. H., Li, B., Zhang, L. & Wang, H. Q. (2008). Locally Linear Discriminant Embedding for Tumor Classification. ICIC, LNAI, 5227, 1093-1100.

[15] Hadid, & PietikÄainen, M. (2003). Efficient locally linear embeddings of imperfect manifolds. In Proceedings of the Third International Conference on Machine Learning and Data Mining in Pattern Recognition, 188-201, Leipzig, Germany, 5-7.

[16] Zhang, Z. & Zha, H. (2003). Local linear smoothing for nonlinear manifold learning. CSE-03-003, Department of Computer Science and Engineering, Pennsylvania State University, University Park, PA, USA.

[17] Hong Chang, (2006). Dit-YanYeung. Robust locally linear embedding, Pattern Recognition, 39, 1053-1065.

[18] Hein, M. & Maier, M. (2006). Manifold denoising. Advances in NIPS 20, Cambridge, MA, 561-568.

[19] Zhang, Z. Y. & Wang, J. (2007). MLLE: Modified locally linear embedding using multiple weights, in: B. Scholkopf, J.C. Platt, T. Hoffman (Eds.), Advances in Neural Information Processing Systems, MIT Press, Cambridge, MA, 19, 171-181.

[20] Yin, J., Hua, D. & Zhou, Z. (2008). Noisy manifold learning using neighborhood smoothing embedding. Pattern Recognition Letters, 29, 1613-1620.

[21] Pan, Y. & Sam, S. G. (2009). Mamun Abdullah Al, Weighted locally linear embedding for dimension reduction. Pattern Recognition, 42, 798- 811.

[22] Zhang, J., Shen, H. & Zhou, Z. H. (2004). Unified Locally Linear Embedding and Linear Discriminant Analysis Algorithm (ULLELDA) for Face Recognition. Lecture Notes in Computer Science, Springer, 3338, 296-304.

[23] Kim, T. K. & Kittler, J. (2005). Locally linear discriminant analysis for multimodally distributed classes for face recognition with a single model image. Pattern Analysis and Machine Intelligence, IEEE Transactions, 27(3), 318-327.

[24] Ridder, D. D., Kouropteva, O., Okun, O., Pietikainen, M. & Duin, R. P. W. (2003). Supervised locally linear embedding. in: Proc. Joint Int. Conf. ICANN/ICONIP2003, in: Lecture Notes in Computer Science, 2714, Springer Verlag, Berlin, Heidelberg, New York, 333-341.

[25] Horn, R. A. & Johnson, C. R. (1999). Matrix Analysis. Cambridge University Press, Cambridge,UK.

[26] Zheng , C. H., Li, B., Zhang, L. & Wang, H. Q. (2008). Locally Linear Discriminant Embedding for Tumor Classification. ICIC, LNAI 5227 1093-1100.

[27] Li, B. & Huang, D. S. (2008). Locally linear discriminant embedding: An efficient method for face recognition, Pattern Recognition, 41(12), 3813-3821.

[28] Khan, J., Wei, J. & Ringner, M. (2001). Classification and diagnostic prediction of cancers using gene expression profiling and artificial neural networks. Nature Medicine, 7, 673-679.

[29] Golub, T. R., Slonim, D. K. & Tamayo, P. (1999). et al, Molecular classification of cancer: Class discovery and class prediction by gene expression monitoring. Science, 286, 531-537.

[30] Nutt, C. L., Mani, D. R. & Betensky, R. A. (2003). et al, Gene expression-based classification of malignant gliomas correlates better with survival than histological classification. Cancer Research, 63(7), 1602-1607.

[31] Van't, L. V., Dai, H., Vijver, M. & He, Y. (2002). et al, Gene expression profiling predicts clinical outcome of breast cancer, Nature, 415, 530-536.

[32] Gordon, G. J., Jensen, R. V. & Hsiao, L. L. (2002). et al, Translation of microarray data into clinically relevant cancer diagnostic tests using gene expression ratios in lung cancer and mesothelioma, Cancer Res, 62, 4968-4976.

[33] Alon, U., Barkai, N. & Notterman, D. A. (1999). et al, Broad pattern of gene expression revealed by clustering analysis of tumor and normal colon tissues probed by oligonucleotide arrays. Proc. Natl. Sci. USA, 96, 6745-6750.

[34] Pomeroy, S. L., Tamayo, P. & Gaasenbeek, M. (2002). et al, Prediction of central nervous system embryonal tumour outcome based on gene expression. Letters to Nature, Nature, 415, 436-442.

[35] Belhumeur, P. N., Hespanha, J. P., Kriegman, D. J. (1997). Eigenfaces vs. Fisherfaces: Recognition Using Class Specific Linear Projection. IEEE Trans. Pattern Analysis and Machine Intelligence, 19(7), 711-720.

[36] Moghaddam, B., Pentland, A. (1997). Probabilistic Visual Learning for Object Representation. IEEE Trans. Pattern Analysis and Machine Intelligence, 19, 696-710.

[37] Boser, B., Guyon, I. & Vapnik, V. (1992). A training algorithm for optimal margin classifiers. In Proc. of 5th Annual ACM Workshop on Computational Learning Theory, 144-152.

[38] Huang, H. L. & Chang, F. L. (2007). ESVM: Evolutionary support vector machine for automatic feature selection and classification of microarray data. *BioSystems*, **90**, 516-528.

In: Biometrics: Theory, Applications and Issues ISBN: 978-1-61728-765-7
Editor: Ellen R. Nichols, pp. 57-80

Chapter 3

FUSION APPROACH FOR IMPROVING THE PERFORMANCE IN VOICE-BIOMETRICS

Di Liu[1,2,*], Siu-Yeung Cho[2], Dong-mei Sun[1] and Zheng-ding Qiu[1]

[1] Institute of Information Science, Beijing Jiaotong University, Shang Yuan Cun 3, Beijing, China,10044

[2] Forensics and Security Laboratory, School of Computer Engineering, Nanyang Technological University, Blk NS1-05-01A, Nanyang Avenue, Singapore, 639798

Abstract

Voice biometrics, also called speaker recognition, is the process of determining who spoke in a recorded utterance. This technique is widely used in many areas e.g., access management, access control, and forensic detection. On the constraint of the sole feature as input pattern, either low level acoustic feature e.g., Mel Frequency Cepstral Coefficients, Linear Predictive Coefficients or high level feature, e.g., phonetic, voice biometrics have been researched over several decades in the community of speech recognition including many sophisticated approaches, e.g., Gaussian Mixture Model, Hidden Markov Model, Support Vector Machine etc. However, a bottleneck to improve performance came into the existence by only using one kind of features. In order to break through it, the fusion approach is introduced into voice biometrics. The objective of this paper is to show the rationale behind of using fusion methods. At the point of view of

[*] E-mail address: liud8310@gmail.com. (Corresponding author)

biometrics, it systematically classifies the existing approaches into three fusion levels, feature level, matching-score level, and decision-making level. After descriptions of the fundamental basis, each level fusion technique will be described. Then several experimental results will be presented to show the effectiveness of the performance of the fusion techniques.

1. Introduction

Nowadays information fusion technique has been widely developed in the community of voice biometrics, besides several unimodal speaker recognition algorithms. Speaker recognition also called voice biometrics usually refers to the process of determining who spoke a recorded utterance. This process may be accomplished by humans alone, who compare a spoken exemplar with the voices of individuals, or by computers, which are programmed to identify similarities in speech patterns. The aim of fusion to speaker recognition is to obtain as much intra-class discriminative information as possible by means of fusion techniques from two or more kinds of input patterns, for examples, short-term spectral features [1], voice source features [2], prosodic features [3] or high-level features [4] to improve performance of the speaker recognition system. Speaker recognition is categorized to two classes [5]. One is **text-dependent speaker recognition** which clients are required to speak specific utterances. This technique requires all the users cooperative. The other is **text-independent speaker recognition** which is suitable for those who are no cooperative. It is more challenging and more widely used in real-world situations e.g., Forensics identification [6]. Forensic Speaker Identification is one of the most important, challenging, but perhaps least well understood applications of speaker recognition. It is aimed at an application area in which criminal intent occurs. This may involve espionage, blackmail, threats and warnings, suspected terrorist communications, etc. Civil matters may hinge on identifying an unknown speaker, as in cases of harassing phone calls that are recorded. To this end, the fusion techniques of the text-independent speaker recognition are mainly discussed to provide a better availability to the issues related to public security than unimodal system. The fusion systems make use of a large amount of speaker discriminative information to yield higher accuracies and robustness than ever. Besides keeping discriminative information, inevitably the systems keep conflict information which may deteriorate the accuracy and redundant information that lead to a high time cost. The conflict information of two or more features has a bad influence on performance, or even yields a worse fused outcome than unimodal. The redundant information is hardly good for the classification procedure, but makes the systems waste a lot of time. Therefore, how to minimize these information is a key to a reasonable performance. Most of the fusion schemes manage to decrease an influence on them.

In the perspective of biometrics [7], four components can be divided as data acquisition, feature extraction, matching score, and decision-making. The data acquisition refers to a procedure that uses a sensor to acquire raw data, for example, utilizes a camera or webcam to acquire images of frontal face in different scenarios, such as controlled condition, noise condition, or complex background condition. The feature extraction is to make use of some algorithms to yield features that stands for clients' intra discriminative information at a large extent. At the same time, these features can be separated easily from that of the other classes. The matching score is based on the features into well trained user models in enrollment [8] to get a group of similarities between templates and probe patterns. After obtaining the matching scores, the decision-making will be conducted. This procedure proposes a functionality of identifying whether claimed user with the input pattern (feature) is genuine or impostor [9] during the task of verification [10]. It categories user with the input pattern to a class in the task of identification [11]. Traditionally, information fusion techniques in biometrics are embedded into each component respectively. For examples, making a fusion in the data acquisition is called **data level fusion** [12]; proposing a fusion scheme at the feature extraction is referred as to **feature level fusion** [13]; figuring out a fused matching score vector or a summed score in the matching score is named **matching-score level fusion** [14]; fusing at the decision-making level is called **decision-making level fusion** [15]. All of them are also applicable in speaker recognition. In text-independent speaker recognition, the most common-used fusion technique is matching-score level fusion. Decision-making fusion is mainly used for kernel fusion i.e., Multiple Kernel Learning (MKL) [16]. In contrast with these two fusion techniques, feature level fusion and data level fusion are rarely employed in speaker recognition. In this review paper, feature level fusion, matching score level fusion and decision-making fusion in text-independent speaker recognition are mainly discussed. After introducing the variety of kinds fusion, a series of evaluations used NIST 2001 corpus [74] are presented to demonstrate the advantage of fusion techniques.

The rest of paper is organized as follows: Section 2 introduces the fundamental fusion techniques, for instance, a categorization at three different fusion levels will be summarized. Section 3 and section 4 discuss feature level fusion and matching score fusion. A kernel fusion also called Multiple Kernel Learning (MKL) at decision-making level is presented in section 5 and the experiments shows the fusion technique is superior to the unimodal in section 6. At last, a conclusion is provided at the last section.

2. Fundamentals of Fusion

Figure 1 shows the components of different levels at a typical speaker recognition system. Unlike image-based biometrics, three main levels are considered, including the feature level, the matching-score level, and the decision-making level. Very few of the data level fusion have been discussed in this field except sub-band fusion [27] that categories into the matching-score level fusion in this paper. As shown in figure 1, the feature level is defined at the component of feature extraction where fusion techniques can be embedded into this component. The matching-score level is defined at the score normalization, more explicitly, fusion techniques can occur prior to this component. Also, a fusion can be made after a score normalization method, e.g. Z-norm [25] or T-norm [26]. Decision-making level is defined at the decision module, and Support Vector Machine (SVM) is commonly employed as the classifier for kernel fusion at this level [22].

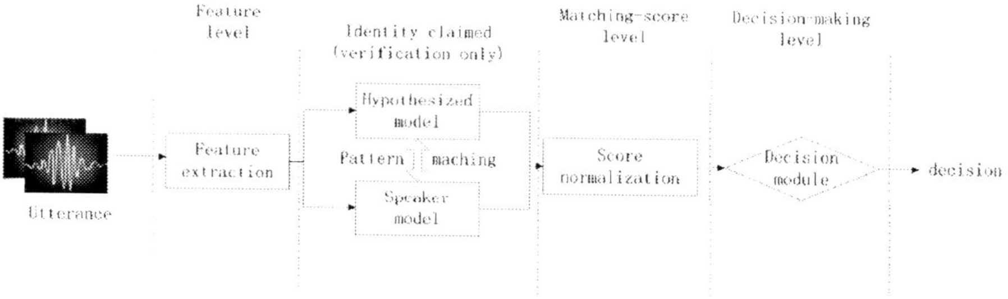

Figure 1. Different fusion levels of a typical automatic speaker recognition. For the sake of simplicity, we here omit the procedure of speaker enrollment. The fusion techniques in speaker recognition can be embedded into each level.

For the feature level fusion, two different features e.g., MFCC [17] and LPCC [18] are fused as a form of feature vector. The fused vectors are used for training of the speaker models. This fusion technique is seldom used in speaker recognition. For the matching-score level fusion, three main sub-categories can be classified, namely **arithmetic fusion**, **vector fusion** and a special case called **self linear combination**. The arithmetic fusion refers to a scheme that sums or multiplies by two or more scores together as a fused score. For a specified Gaussian Mixture Model (GMM) [19] based speaker model, the log likelihood ratios (LLR) [20] produced by the model are used as the scores. Suppose that two different features are extracted as inputs to the model, e.g., MFCC and residual phase [21], two kinds of scores can be yield respectively. The arithmetic fusion technique sums these two scores together to derive a resultant LLR. Also, there are other rules, e.g., Max and Min rules that can be used. Moreover, another early used method called vector fusion technique which makes different LLRs as a form of vector rather than the sum or product. Usually, a classifier e.g., SVM and cross validation technique [23] are utilized to train and test these fused vectors. For example, GMM-SVM system [24] is a typical vector fusion. In addition, a special case of self linear combination of LLR will be discussed in this review paper. For decision-making level fusion, it is mainly concerned with kernel fusion, i.e., the techniques of multiply kernel learning in speaker recognition.

Table 1. A statistics of fusion techniques used in speaker recognition

Authors	Fusion method	Authors	Fusion method
Higgins et al. [27]	Matching-score fusion	Kajarekar et al. [62]	Matching-score fusion
Vale et al. [47]	Matching-score fusion	Charbuillet et al. [63]	Matching-score fusion
Vale et al. [48]	Matching-score fusion	Murty et al. [65]	Matching-score fusion
Chakroborty et al. [49]	Matching-score fusion	Long et al. [66]	Matching-score fusion
Hu et al. [50]	Matching-score fusion	Wang et al.[67]	Matching-score fusion
Jin et al. [52]	Matching-score fusion	Dehak et al. [68]	Decision-making fusion
Wang et al. [54]	Feature level fusion	Deng et al. [58]	Matching-score fusion
Garcia-Romero et al. [55]	Matching-score fusion	Bengio et al. [59]	Matching-score fusion
Zheng et al. [56]	Matching-score fusion	Nosratighods et al. [60]	Matching-score fusion
Memon et al. [57]	Matching-score fusion	Ma et al. [32]	Feature level fusion
Islam et al. [61]	Matching-score fusion	Longworth et al. [43]	Decision-making fusion

According to the three levels defined, recent works of fusion techniques are listed in Table 1. As shown in the table, the matching-score fusion techniques are popular. Two other fusion techniques, the feature level fusion and the decision-making level fusion are rarely used. It is quite different from that of image-based biometrics fusion, which the feature level fusion is common used.

Moreover, a hierarchy of fusion levels and fusion categories is shown as figure 2. Note that the term "level" in figure 2 is different from Tomi and Li's "level" [2]. In their work, the term "level" is referred as to variety of aspects features in speaker recognition. They divided the features into five typical classes in the field of speaker recognition, namely short-term, voice source, spectral-temporal, prosodic feature, and high-level feature. For example, the short-term feature, e.g., MFCC can be categorized into low level feature. And the idiolect feature is considered as high level feature. Their "levels" are just defined in only the component of feature extraction. On the contrary, the "level" in this paper denotes different fusion levels at three different components in figure 1, i.e., the component of feature extraction, matching-score level, and decision level. The decision-making level is defined as the high fusion level, and feature level as the low fusion level.

As shown in figure 2, it is evident that matching-score fusion is the most popular chosen approaches, and it has more subcategories than other levels. At this level, the arithmetic fusion is usually employed, especially its subcategory named weighted sum rule, also called linear weighted combination, e.g., [47-50, 52].

3. Feature Level Fusion in Speaker Recognition

Feature level fusion is referred as to computing a feature vector via the extraction from sensors. It is meant that each feature vector extracted from its corresponding sensor concatenates into a single vector with a high-dimensionality, and then a dimensionality reduction technique will be employed for the new fused vector. This fusion technique is widely used in the image-based biometrics [28-31], but very few for voice biometrics. Ma et al. [32] proposed a feature level fusion algorithm named Further Feature Extraction which fused among weighted LPCC, weighted MFCC and weighted ΔMFCC [33] to a high dimensional matrix, then used PCA technique [34] to conduct feature selection. Therefore, the principle components of the fused feature matrix would be used as the feature representation. Here, the aim of PCA was not only to implement an optimal feature selection, but also to play a role on the dimensionality reduction in case of the curse of dimensionality [35]. Wang and Wang [54] combined the features of DMEGB and ΔDMEGB to accomplish the performance evaluation, which DMEGB is referred as to the difference between mean energy in generalized bark of the fractional cosine transform (FRCT) and that of the fractional sine transform (FRST) [69]. The equation can be expressed as follows

$$\Delta E^{p} = E_{c}^{p} - E_{s}^{p} \tag{1}$$

where E_{c}^{p} and E_{s}^{p} denote MEGB vectors using FRCT and FRST respectively. Δ DMEGB is the difference of DMEGB of FRCT and FRST which is described as follows:

$$\Delta\Delta E^{p} = \Delta E_{c}^{p_1} - \Delta E_{s}^{p_2} \tag{2}$$

where $\Delta E_{c}^{p_1}$ and $\Delta E_{c}^{p_2}$ indicated DMEGB feature vector of FRCT and FRST in primary p_1 th order fractional domain and the one of secondary p_2 th order fractional domain respectively.

The feature level fusion is rarely used in speaker recognition. Although it can preserve intra-discriminative information at a large extent in image-based biometrics, a series of complicated procedure have to be implemented.

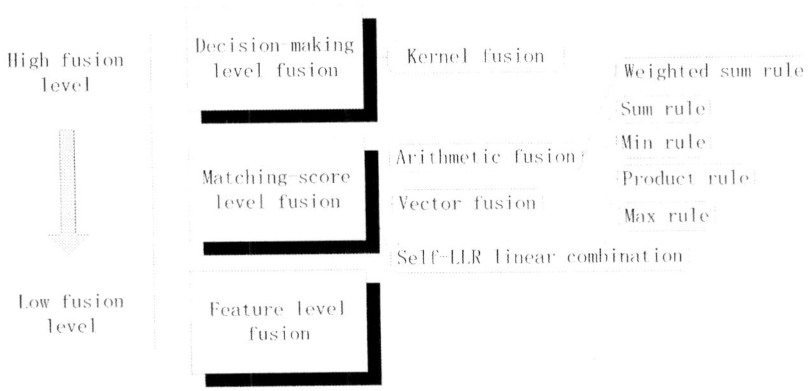

Figure 2. Hierarchy of fusion levels and fusion categories in automatic speaker recognition.

4. Matching-Score Level Fusion in Speaker Recognition

This section presents the matching-score level fusion techniques in automatic speaker recognition. As mentioned in section 2, this level fusion can be categorized into the arithmetic fusion, vector fusion and self linear combination.

4.1. Arithmetic Fusion

Arithmetic fusion emerges prior to the vector fusion. Its definition is: for a statistic-based speaker model, e.g. HMM [36] or GMM, two kinds of features A and B are extracted by the speaker recognition system. Then two corresponding LLRs $\log p_A(x|\theta)$ and $\log p_B(x|\theta)$ are derived by the same speaker model, where $p_A(x|\theta)$, $p_B(x|\theta)$ are probability distribution functions (PDF) of features A and B, and θ is a set of parameters of the model. The arithmetic fusion is to make a fusion between features A and B as a sum of

$$\log p_{fusion}(x|\theta) = \log p_A(x|\theta) + \log p_B(x|\theta) \tag{3}$$

or a product of

$$\log p_{fusion}(x|\theta) = \log p_A(x|\theta) \cdot \log p_B(x|\theta) \tag{4}$$

For multiple features, the fused score is defined as either:

$$\log p_{fusion}(x|\theta) = \sum_{i=1}^{n} \log p_i(x|\theta) \tag{5}$$

or

$$\log p_{fusion}(x|\theta) = \prod_{i=1}^{n} \log p_i(x|\theta) \tag{6}$$

where n is the number of features. The sum of likelihood ratios method is more widely employed in arithmetic algorithm than the product of likelihood ratios. It is worth to note that an extension of Eq. (5), the method of the weighted sum also called **linear weighted combination** [49, 57, 60, 61, 63] is frequently employed in the recent publications:

$$\log p_{fusion}(x\,|\,\theta) = \sum_{i=1}^{n} \alpha_i \log p_i(x\,|\,\theta), \qquad \sum_{i=1}^{n} \alpha_i = 1 \tag{7}$$

where α_i is the weight for LLRs of each features. Most of the works chose the linear weighted combination, whereas there are Max, Min and Median rules are used for the other works [50].

Early the arithmetic fusion techniques were employed in subband algorithms [46]. Higgins et al. [27] proposed an algorithm of information fusion for subband HMM model speaker recognition by a simple sum of likelihood ratios. Vale et al. [47] conducted a linear weighted sum fusion to investigate the robustness of the algorithm in color noise environments, e.g. car noise, factory noise, or white noise. Based on the weights determined by the energy of subband signal, an adaptive weighting method was proposed to improve the robustness under the noise scenarios. This approach was based on the computation of a basis for the null space to implement an adaptive combination of subband signal responses [48]. Chakroborty et al. [49] made a linear weighted combination of two features MFCC and Inverted Mel Frequency Cepstral Coefficient (IMFCC) [72] that is derived from the IMFCC triangular filter bank. This filter bank is actually a reversal of MFCC filter bank in the procedure of producing MFCC, which is the only difference between MFCC and IMFCC. IMFCC can capture the speaker specific information from the high frequency range of the spectrum on the opposite of the MFCC extraction from the low frequency range. Therefore, the IMFCC feature provides complementary information to low frequency sensitive MFCC. Also, Memon et al. [57] investigated the performance of between MFCC, IMFCC, and linear weighted combination fusion. As a result, the fusion technique outperforms the single feature of MFCC or IMFCC. Long et al. [66] used a linear weight combination method to fuse acoustic feature and high level prosodic feature. In their work, the high level prosodic was divided into two types, the prosodic related features (log pitch and log energy) and dynamic parameters, i.e., the first and second derivatives of pitch and energy. Murty et al. [65] considered the feature of residual phase is a complementary to low level spectral-based feature MFCC. And they fused two features by means of the linear weighted combination. Charbuillet et al. [63] made a weighted sum of LLRs with optimal representations of LPCC by genetic algorithm. The use of genetic algorithm was to find the best values of the system's parameters in order to obtain a reasonable performance. Islam et al. [61] conducted a modification of the weighted sum fusion, which weight is calculated as

$$S_{fuse} = ((S_{HMM} - S_{AHS}) \times \omega) + S_{AHS} \quad , \omega = \frac{M_{AHS}}{M_{AHS} + M_{HMM}} \tag{8}$$

where M_{AHS} and M_{HMM} are the means of normalized scores from Arithmetic Harmonic Sphericity Measure (AHS) and HMM. Nosratighods et al. [60] used the FoCal software [73] to make a logistic regression fusion of Frequency Modulation (FM) and MFCC features. Hu et al. [50] did a linear weighted sum method to fuse the scores from the utterance signal with silence and without silence. Zheng et al. [56] proposed a weighted sum between score from MFCC and that of Wavelet Octave Coefficients of Residues (WOCOR). The feature WOCOR is used to supplement the conventional spectral-based feature, e.g., MFCC. Jin et al. [52] used five scores to make a linear weighted combination to improve the performance. Wang et al. [67] used a linear coupling method to fuse 3 kinds of likelihood scores of HMM/GMM, which integrated phase information and acoustical feature of MFCC.

4.2. Vector Fusion

Besides the arithmetic fusion, many researchers focused on fusing multiple scores as a form of vector to for classification e.g., linear regression [37], Neutral Network [38], or SVM classifier. For two features A, B, $\log p_A(x|\theta), \log p_B(x|\theta)$ are LLRs with respect to A and B respectively. The vector fusion technique makes them as a vector

$$f = [\log p_A(x|\theta) \quad \log p_B(x|\theta)] \tag{11}$$

Then this kind of fused vector is used for training and testing for classification task. For multiple features, the fused vector is expressed as

$$f = [\log p_1(x|\theta) \quad \log p_2(x|\theta) \quad \cdots \quad \log p_n(x|\theta)] , \tag{12}$$

where n is the number of features. To make the performance of the classifier stable, n-folder cross validation is commonly used. That is, all the fused vectors built can be divided into n partitions. One partition is viewed as a testing set, and the rest of them i.e., n-1 partitions as training sets. The system runs n times which each partition is training set in turn. Generally speaking, this kind of fusion technique is embedded on the SVM classifier [39]. Garcia-Romero [55] et al. proposed a vector fusion for a classification of SVM followed with the idea of [69]. This is a typical GMM-SVM based vector fusion technique. Kajarekar et al. [62] proposed a linear weighted sum score of four systems after SVM. The features of 4 subsystems are the first PCs of FCA, PCs with zero eigenvalue of FCA, the first N PCs of PCA, and that of with zero eigenvalue of PCA. Dehak et al. [68] proposed a SVM combined fusion between the speaker factor space and the common factor space to yield a higher accuracy.

4.3. A Special Case at Matching-Score Fusion: Linear Combination of Self-LLR

Linear combination of self-LLR is referred to as using an optimal linear combination method to represent the relationship between speaker LLR and UBM or Cohort LLR [26], to maximize speaker intra-class specific information. Based on the GMM-UBM model, LLR is computed as

$$LLR = \log p(X \mid S_i) - \log p(X \mid \Omega) \tag{13}$$

where $\log p(X \mid \Omega)$ is the log likelihood ratio produced by all users in the corpus. The technique of linear regression fusion is to determine a group of optimal parameters of each term above to maximize the speaker discriminative information. Generally speaking, Eq. (13) is modified in this technique as follows:

$$y = a \log p(X \mid S_i) - b \log p(X \mid \Omega) + c \tag{14}$$

where a, b, c are the parameters for linear combination of the final score output. Bengio et al. [59] used with this equation to yield an optimal solution of a, b, c by linear regression. Deng et al. [58] employed the linear weighted combination among speaker, UBM and Cohort LLR to enlarge speaker intra-class discriminative information.

To sum up, this level fusion techniques are widely utilized in speaker recognition. Compared with other two levels fusion, it can gain more reliability. And a statistics table of recent algorithms fused in matching-score level is showed on table 2.

5. Decision-Making Level Fusion in Speaker Recognition

This section describes fusion technique at the level of decision-making. This kind of fusion is rarely employed in speaker recognition. In this paper we mainly consider the kernel fusion. The kernel fusion also called kernel combination is to make a fusion among different kernels in some kernel based methods. After deriving LLRs, some kernel-based classifiers is used, e.g., SVM. SVM is a binary classifier that manages to make a maximum separating margin between positive samples and negative samples. For a kernel-based method, its aim is to predict a value of an unknown label [42]:

$$\hat{S} = sign(\sum_i S_i \lambda_i K(X_i, X)) \tag{15}$$

where $X_{i=1\cdots n}$ are training sample (vectors), $S_{i=1\cdots n}$ the labels with respect to X_i, $\lambda_{i=1\cdots n}$ the weights with respect to X_i, and $K(X_i, X)$ is the kernel function. In the context of SVM, the solution of the learning problem is of the form [40]:

$$f(x) = \sum_{i=1}^{l} \alpha_i y_i K(x, x_i) + b^* \tag{16}$$

where α_i is a weight as λ_i of Eq. (15), y_i is either 1 or -1 as a binary classifier, b^* is a coefficient to be learned from examples. $K(X_i, X)$ is also a kernel function. The kernel fusion techniques are implemented at this term. In such cases, a common approach is to consider that the kernel $K(X_i, X)$ can be a convex linear combination of other basis kernels:

$$K(X_i, X) = \sum_{k=1}^{m} d_k K_k(X_i, X), d_k \geq 0, \sum_{k=1}^{m} d_k = 1 \quad (17)$$

where $K_k(X_i, X)$ is one kind of kernel, e.g., RBF, linear, or polynomial kernel [41]. Eq.(17) is a basic formluation of typical kernel fusion technique. Dehak et al. [64] used this kind of kernel fusion method to fuse between Generalized Linear Discriminate Sequence (GLDS) kernel [70] and GMM-SVM Linear kernel [71]. Longworth et al. [43] employed multiple kernels learning technique to find out a group of optimal weights for each fused kernel according to Eq. (17). Dehak et al. [68] made a kernel fusion between two new kernel functions

$$k(x, x') = x^t W^{-1} x' \quad (18)$$

and

$$k(x, x') = \frac{x^t W^{-1} x'}{x^t W^{-1} x x' W^{-1} x'} \quad (19)$$

by linear combination.

Table 2. A statistics table of recent algorithms fused in matching-score level

Authors	Algorithm/Content	Fusion category
Higgins et al. [45]	A sum fusion of HMM LLRs of each subband signal.	Arithmetic fusion (sum)
Vale et al. [47]	A weighted sum of GMM LLRs of subband signals, weighting by the energies of subband signals.	Arithmetic fusion (weighted sum)
Vale et al. [48]	An adaptive weight sum of LLRs method to improve the robustness in colored noise scenarios.	Arithmetic fusion (weighted sum)
Chakroborty et al. [49]	A linear weighted combination of LLRs of features MFCC and IMFCC.	Arithmetic fusion (weighted sum)
Hu et al. [50]	A linear weighted fusion between the score of the utterance signal with silence and without slience.	Arithmetic fusion (weighted sum)
Jin et al. [52]	A weighted sum of 5 kinds of scores.	Arithmetic fusion (weighted sum)

Table 2. Continued

Authors	Algorithm/Content	Fusion category
Garcia-Romero et al. [55]	A score combination of vector for SVM classification.	Vector fusion
Zheng et al. [56]	A linear weighted combination of the scores of MFCC and WOCOR.	Arithmetic fusion (weighted sum)
Memon et al. [57]	A linear weighted sum of LLRs of features MFCC and IMFCC.	Arithmetic fusion (weighted sum)
Deng et al. [58]	A weighted linear combination for the representation of LLR among speaker GMM model, UBM model, and Cohort model.	Self-LLR fusion
Bengio et al. [59]	A linear regression method to obtain an optimal linear combination of LLR scores for each speaker	Self-LLR fusion
Nosratighods et al. [60]	A logistic regression fusion of FM and MFCC.	Arithmetic fusion (weighted sum)
Islam et al. [61]	A modification of linear weighted combination by AHS and HMM models.	Arithmetic fusion (weighted sum)
Kajarekar et al. [62]	A linear weighted sum of score of four systems, first PCs of PCA, PCs with zero eigenvalue of FCA, first N PCs of PCA, and that of with zero eigenvalue of PCA.	Vector fusion
Charbuillet et al. [63]	A weighted sum of LLRs of the optimal representations of LPCC by genetic algorithm.	Arithmetic fusion (weighted sum)
Murty et al. [65]	A weighted sum of LLRs from two features MFCC and residual phase, which residual phase can provide complementary information to MFCC.	Arithmetic fusion (weighted sum)
Long et al. [66]	A linear weight combination method to fuse acoustic feature and high level prosodic feature.	Arithmetic fusion (weighted sum)
Wang et al.[67]	A linear coupling method with 3 kinds of LLR scores of GMM/ HMM model.	Arithmetic fusion (weighted sum)
Dehak et al. [68]	A SVM combined fusion between the speaker factor space and the common factor space.	Vector fusion

To sum up, this section mainly introduces the idea of decision-making level fusion, i.e., kernel level fusion. However, this aspect of fusion techniques is less common-used than the matching-score level in speaker recognition.

6. Evaluation

This section mainly demonstrates the performance of different fusion techniques for speaker recognition. As mentioned in the previous sections, here we discuss two main level fusion techniques, i.e., the feature level fusion and the matching-score level fusion. Three evaluations will be presented as below. For these evaluations, we use NIST2001corpus [74].

The first evaluation was conducted with the fusion between MFCC and residual phase. MFCC is a common-used spectral feature in speaker recognition, whereas residual phase is complementary to the MFCC feature during the processing the fusion [65]. Therefore, the performance of their fusion results is being tested.

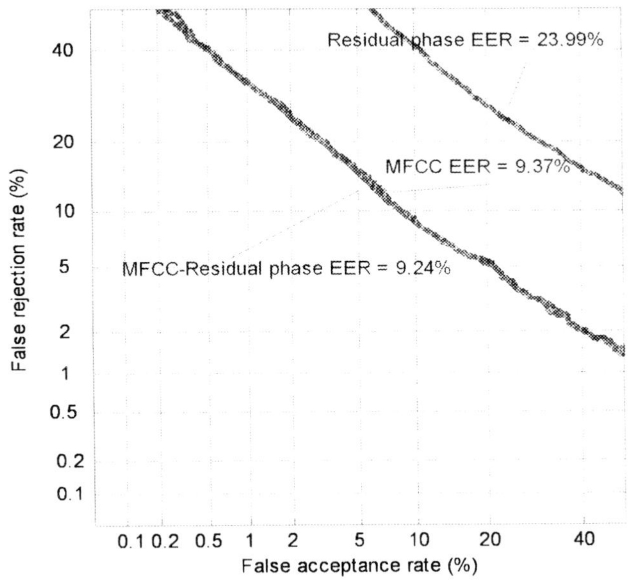

Figure 3. A comparison between MFCC-residual phase fusion and their sole features.

As seen from figure 3, it is found that MFCC baseline method is able to achieve Equal Error Rate (EER) as 9.37% whereas the residual phase achieved as 23.99%. The fusion between these two features obtains EER 9.24%, which is better than both of only using single feature. Here the fusion technique that we adopted is the vector fusion which makes two feature LLRs as a 1×2 vector. It obtains intra-class discriminative information from two separate features. Two scores with this kind of information concatenate as a vector for training and testing. This form is useful to improve the performance as shown on the fusion result.

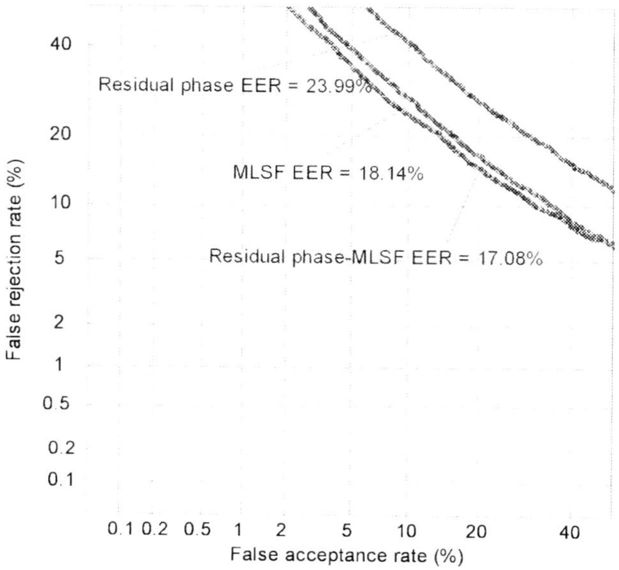

Figure 4. A comparison among residual phase, MLSF and fusion.

As seen from figure 4, the EER obtained from residual phase and MLSF [75] are 23.99% and 18.14% respectively. After the fusion between the residual phase and MLSF, the EER is reduced to 17.08%. It means that additional discriminative information can be acquired to reduce the error rate.

However, it does not always obtain a promising performance using fusion. For example, on the same evaluation condition, figure 5 shows that the fusion between MFCC and MLSF that is not better than that by MFCC baseline in which its EER is 9.37%. The outcome of the fusion is 11.20%, which means the conflict information in features deteriorates the fused result, and hence affecting the effectiveness of the discriminative information. In this case, the conflict information has a larger influence on the fusion to discriminate them. That is a reason why a worse performance was obtained rather than using single MFCC-baseline.

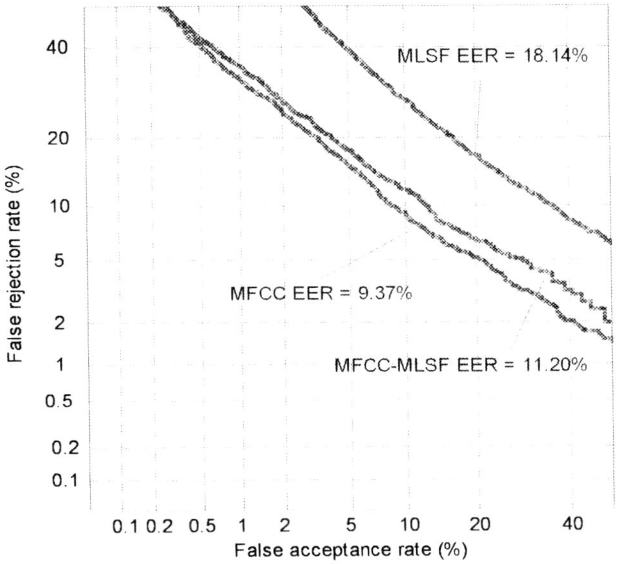

Figure 5. A comparison among MFCC, MLSF, and fusion.

The last experiment is a comparison between the feature level fusion and matching-score level fusion. It aims to demonstrate the feature level fusion that is superior to the matching-score level one. The pH feature is a vector of Hurst parameters obtained from the windowed short-time segments of speech [45]. As shown in figure 6, the single pH4, pH6, pH12 obtained the EER as 32.19%, 33.66%, and 31.75% respectively. Through the vector fusion among pH4, pH6, and pH12, a sub-category of the matching score level fusion, the fused EER is 30.03%. On the contrary, the feature level fusion concatenates these three kinds of pH feature vectors into a fused one which yields 26.30% EER. This comparison implies that the feature level fusion can extract more discriminative information than the matching score level one in speaker recognition, just similar to the image-based biometrics [12]. Also, the fusion of the two features is worse than that of these features, e.g., pH4 and pH6, pH4 and pH12, and pH6 and pH12. This proves that the discriminative information of the three feature fusion is richer than only using two features.

In conclusion, this section shows a series of fusion evaluations. For the first and second illustrative experiments show that fusion results are superior to single features, according to the accumulation of the discriminative information amount. The third one demonstrates the case that the conflict information can affect performance much more than the discriminative information. This case should be avoided. The last one shows the feature level fusion can extract more discriminative information than the matching-score level fusion, to yield a remarkable EER. And if we manage to use some algorithms to represent the original pH4, pH6, and pH12 features, it may get a better result, e.g., PCA or ICA algorithm.

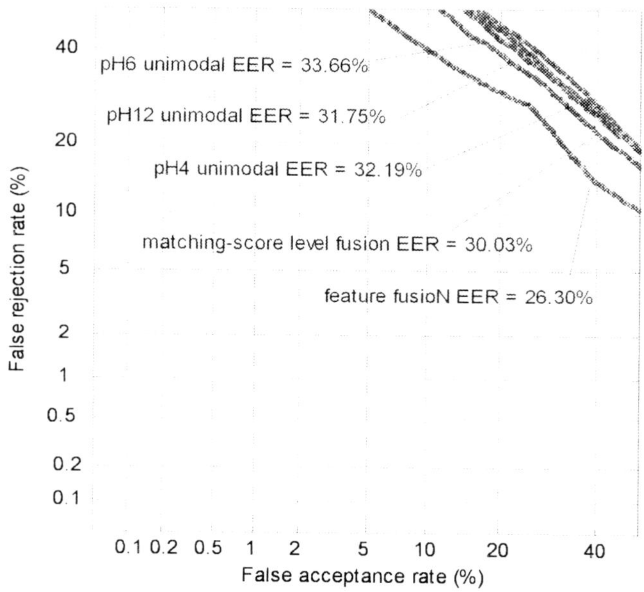

Figure 6. A comparison between the feature level fusion and the matching-score level fusion.

7. Conclusion

This paper presents different fusion techniques for automatic speaker recognition. Based on the similar idea of fusion levels used in image-based biometrics, the fusion techniques used in voice biometrics can be divided into three main categories. They are: the feature level fusion, the matching-score level fusion and the decision-making level fusion. Most of the fusion worked in speaker recognition are employed the matching-score level fusion, especially arithmetic fusion that is the sub-category of the matching-score fusion. Also, vector fusion is another subcategory of this level fusion that usually being used in the GMM-SVM systems. For each level fusion, its corresponding principles and examples have been discussed in this paper as well. In addition, the experiments have shown the advantage of the fusion techniques. The main purpose of the fusion techniques is to improve the overall performance of the front-to-end system.

Acknowledgments

This work is partially funded by Grant No. 60773015 of National Science Foundation of China, Grant No. 20060004007 of Doctoral Program of Higher Education, Grant No. 4102051 of Beijing Natural Science Foundation, and Grant No. 2009JBZ006 of the Fundamental Research Funds for the Central Universities.

References

[1] Claude, C., Farzin, D. & John, S.D. (2002). A review of speech-Based bimodal recognition. *IEEE Transaction on multimedia*, vol.4, no.1, 23-37.

[2] Kinnunen, T., & Li H. (2010). An overview of text-independent speaker recognition: from features to supervectors. *Speech Communication*, vol. 52, no. 1, 12-40.

[3] Long, Y., Ma, B., Li, H., Guo, W., Chug, E. S., & Dai, L. (2009). Explorting prosodic information for speaker recognition. In: *Proceeding of ICASSP* 2009, 4225-4228.

[4] Zhang, S. X., Mak M. W., & Meng, M. (2007). Speaker verification via high-level feature based phonetic-class pronunciation modeling. *IEEE Transaction on Computers*, vol. 56, no. 9, 1189-1198.

[5] Campbell, W. M. (1997). Speaker recognition: A tutorial. Proceeding of The IEEE, vol. 85, no. 9, 1437-1462.

[6] Joseph, P., Campbell, Shen, W., William M., Campbell, Reva S., Jean-François B. & Driss M. (2009). Forensic speaker recognition. *IEEE Signal Processing Magazine*, vol. 26 no.2, 95 -103.

[7] Hong, L., Jain, A. & Pankanti,S. (1999). Can Multibiometric Improve Performance? In: *Proceeding of AutoID'99 Summit*, 59-64.

[8] Reynolds, D. A., Thomas, F., Quatieri, & Dunn, R. D. (2001). Speaker verification using adapted Gaussian Mixture Models. *Digital signal processing*, 19-41.

[9] Jain, A., Nandakumar, K. & Ross, A. (2005). Score normalization in multimodal biometric systems. *Pattern Recognition*, vol. 38, no. 12, 2270-2285.

[10] Mary, A. K., Walter, D., Andrews, J. P., Campbell, & Jaime, Hernandez-Cordero. (2001). Phonetic speaker recognition. In: *Proceeding of Eurospeech, Aalborg*, 1557-1661.

[11] Jain, A., Bolle, R. & Pankanti, S. In: *Book of Biometrics: Personal Identification in Networked Society*. Kluwer Academic Publishers.

[12] Jain, A. & Ross, A. (2004). *Multibiometric Systems. Communication of the ACM*, vol.47, no. 1, 34-40.

[13] Li, Q., Qiu, Z. & Sun, D. (2006). Feature-level fusion of hand biometrics for personal verification based on kernel PCA. In: *Proceeding of Advance in Biometrics*, 744-750.

[14] Varchol, P., Levicky, D. & Juhar, J. (2008). Multimodal biometric authentication using speech and hand geometry fusion. *Systems, Signals and Image Processing, 15th International Conference on Publication*, 57-60.

[15] Ross, A. & Jain, A. (2003). Information fusion in biometrics. *Pattern Recognition Letters*, vol. 24, no.13, 2115-25.

[16] Alain, Rakotomamonjy, Francis, Bach, Stephane, Canu, & Yves, Grandvalet. (2007). More efficiency in multiple kernel learning. In: *Proceedings of the 24th international conference on Machine learning*, vol.227, 775-782.

[17] Holmes, J., & Holmes, W. (2003). *Speech synthesis and recognition*: Taylor& Fransic Group publisher.

[18] Eriksson, T., Kim, S., Kang, H. G., & Lee, C. ((2005). An information-theoretic perspective on feature selection in speaker recognition. *IEEE signal processing letters*, vol.12 no.7, 500-503.

[19] Chao, Y. H., Tsai, W. H., & Wang, H. M. (2009). Improving GMM-UBM speaker verification using discriminative feedback adaptation. *Computer speech and languages*, vol. 23, 376-388.

[20] Campbell, W. M., Campbell, J. P., Reynolds, D. A., Singer, E. & Torres-Carrasquillo, P. A. (2006). Support vector machines for speaker and language recognition. *Computer speech and language,* **20**, 210-219.

[21] Islam, T. & Kabal, P. (2000). Partial-energy weighted interpolation of linear prediction coefficients. In: *Proceeding of IEEE Workshop Speech Coding* (Delavan, WI), 105-107.

[22] Chistopher, Burges. (1998). A tutorial on support vector machine for pattern recognition. *Data mining and knowledge discovery*, **2**, 121-167.

[23] Hsu, C. W., Chang, C. C., & Lin, C. J. A Practical Guide to Support Vector Classification. Available from: http://www.csie.ntu.edu.tw/~cjlin/papers/guide/guide.pdf.

[24] Campbell, W. M., Sturim, D. E. & Reynolds, D. A. (2006). Support vector machines using GMM supervectors for speaker recognition. *IEEE singal processing letters*, vol 13, no.5, 308-311.

[25] Auckenthaler, R. & Mason, J. (2001). Gaussian selection applied to text-independent speaker verification. In Proc. *Speaker Odyssey*, 83-88.

[26] Auckenthler, R., Carey, M. & Lloyd-Thomas, H. (2000). Score normalization for text-independent speaker verification systems. *Digital signal processing*, vol. 10, 42-54.

[27] Higgins, J. E., Dodd, T. J. & Damper, R. I. (2001). Information fusion for subband-HMM speaker recognition. In: Proceeding of *INNS-IEEE International Joint Conference on Neural Networks, IJCNN'01*, Washington DC. 1504-1509.

[28] Hong, L. & Jain, A. (1998). Integrating faces and fingerprints for personal identification, *IEEE Trans. on Pattern Analysis and Machine*, vol.20, no.12, 1295-1307.

[29] Huang, R., Liu, Q., Lu, H. & Ma, S. (2002). Solving the Sample Size Problem of LDA. In: Proceeding of 16th *International Conference on Pattern Recognition*, 29-32.

[30] Li, Q., Qiu, Z., Sun, D. & Zhang, Y. (2006). Subspace Framework for Feature-Level Fusion with Its Application to Handmetric Verification. In: Proceedings of 8th *International Conference on Signal Processing*, 16-20.

[31] Faraj, M. & Bigun, J. (2007). Audio-visual person authentication using lip-motion from orientation maps. *Pattern Recognition Letters* vol. 28, no. 11, 1368-82.

[32] Ma, Z., Yang, Y., & Wu. Z. (2003). Further feature extraction for speaker recognition. vol.5, 4135-4138.

[33] Wang, J. & Wang, J. (2005). Speaker recognition using features derived from fractional fourier transform. In: *Proceeding of IEEE International Conference on Automatic Identification Advanced Technologies*, 95-100.

[34] Li, Q., Qiu, Z. & Sun, D. (2006). Feature-level fusion of hand biometrics for personal verification based on kernel PCA. In: *Proceeding of Advance in Biometrics*, 744-50.

[35] Bellman, R. E. (1961). Adaptive Control Processes: Princeton University Press, Princeton, NJ.

[36] Minh, N. Do. (2003). Fast approximation of Kullback-Leibler distance for dependence trees and hidden Markov models. *IEEE signal processing letters*, vol.10, no. 4, 115-118.

[37] Siu-Kei Au-Yeung, Man-Hung Siu. (2006). Maximum likelihood linear regression adapatation for the Polynomial segment models. IEEE *Signal Processing Letters*, vol. 13, no. 10, 644-647.
[38] Cho, S.Y. (2008). Probabilistic Based Recursive Model for Adaptive Processing of Data Structure. *Expert Systems with Applications*, vol.32 no.2, 1403-1422.
[39] You, C. H., Lee, K. A., Li, H. (2009). An SVM kernel with GMM-Supervector based on the Bhattacharyya distance for speaker recognition. *IEEE signal processing letters*, vol.16, no.1, 49-52.
[40] Rakotomamonjy, A., Bach, F., Canu, S., & Grandvalet, Y. (2007). More efficiency in multiple kernel learning. In: Proceeding of the 24th *International Conference on Machine learning*, 775-782.
[41] Wan, V., Renals, S. (2002). Evaluation of kernel methods for speaker verification and identification. *Icassp*, 2002, vol.1, 669-672.
[42] Tommi, S., Jaakkola, & David Haussler. (1998). Exploiting generative models in discriminative classifiers. In Advances in Neural *Information Processing Systems*, **11**, MIT Press, 487-493.
[43] Longworth, C. & Gales, M. J. F. (2008). Multiple kernel learning for speaker recognition. ICASSP 2008, 1581-1584.
[44] Berhard, S. (2000). *The kernel trick for distances*, MIT Press, 301-307.
[45] Sant'Ana, R., Coelho, R. & Alcaim, A. (2006) Text-independent speaker reocognition based on the hurst parameter and the multidimensional fractional brownian motion model. IEEE Transaction on Audio, *Speech and Language processing*. vol.14, no.3, 931-940.
[46] Sakka, Z., Kachouri, A., Mezghani, A. & Samet M. (2004). A new method for speech denoising and speaker verification using subband architecture. In: *Proceeding of First International Symposium on Control, Communications and Signal Processing*, 37-40.
[47] Vale, E. E., Cunha, A. & Alcaim, A. (2008). Robust text-independent identification using multiple subband-classifiers in colored noise environment. 15th International Conference on Systems, *Signals and Image Processing*, 2008, 275-278.
[48] Vale, E. E. & Alcaim, A. (2008). Adaptive weighting of subband-classifier responses for robust text-independent speaker recognition. *Electronics letters*, vol.44, no.21, 1280-1282.
[49] Chakroborty, S. & Saha, G. (2009). Improved text-independent speaker identification using fused MFCC and IMFCC feature sets based on Gaussian filter. *International journal of signal processing*, vol.5, no. 1, 11-19.

[50] Hu, R. & Damper R. I. (2005). Fusion of two classifiers for speaker identification: removing and not removing silence. In: *Proceeding of 7th international conference on information fusion*, 429-436.

[51] Hu, R., & Damper R. I. (2001). A combination between VQ and covariance matrices for speaker recognition. *ICASSP* 2001, vol. 1, 453-456.

[52] Qin Jin, Jiri Navratil, Douglas A. Reynolds, Joseph P. Campbell, Walter D. Andrews, & Joy S. Abramson. (2003). Combining cross-stream and time dimensions in phonetic speaker recognition. *ICASSP*, 2003, 800-803.

[53] Kajarekar, S., Ferrer, L., Venkataraman, A., Sonmez, K., Shriberg, E., Stolcke, A., Bratt, H., Rao, R. & Gadde. (2003). Speaker recognition using prosodic and lexical features. *ARSU*, 19-24.

[54] Jinfang Wang, & Jinbao Wang. (2005). Feature extraction method of fractional cosine and sine transform for speaker recognition. In: Proceedings of the IEEE International Conference on Electronics, *Circuits, and Systems*, 4-8.

[55] Garcia-Romero, D., Fierrez-Aguilar, J., Gonzalez-Rodriguez, J. & Ortega-Garcia, J. (2003). Support vector machine fusion of idiolectal and acoustic speaker information in Spanish conversational speech. *Icassp*, vol. 2, 229-232.

[56] Nengheng Zheng, Tan Lee, P. C. Ching. (2007). Integration of complementary acoustic features for speaker recognition. *IEEE signal processing letters*, vol.14, no.3, 181-184.

[57] Sheeraz Memon, Margaret Lech, & Ling He. (2009). Using information theoretic vector quantization for inverted MFCC based speaker verification. In: *Proceeding of 2nd International Conference on Computer, Control and Communication*, 1-5.

[58] Haojiang Deng, Limin Du, & Hongjie Wan. (2004). Combination of likelihood scores using linear and SVM approaches for text-independent speaker verification. *ICSP* 04, 2261-2264.

[59] Samy Bengio, & Johnny Mariethoz. (2007). Learning the decision function for speaker verification. *ICASS*P, 425-428.

[60] Nosratighods, M., Thiruvaran, T., Epps, J., Ambikairajah, E., Ma, B. & Li, H. (2009). Evaluation of a fused FM and cepstral-based speaker recognition system on the NIST 2008 SRE. *ICASSP*, 4233-4236.

[61] Islam, T., Mangayyagari, S. & Sankar, R. (2007). Enhanced speaker recognition based on score level fusion of AHS and HMM. *IEEE Southeast Con*, 14-19.

[62] Kajarekar, S. S. (2005). Four weightings and a fusion: a cepstral-SVM system for speaker recognition. *AUSR*, 17-22.

[63] Charbuillet, C., Gas, B., Chetouani, M. & Zarader, J. L. (2007). Complementary features for speaker verification based on genetic algorithms. *ICASSP* 2007, vol. IV 285-288.

[64] Dehak, R., Dehak, N., Kenny, P. & Dumouchel, P. (2008). Kernel combination for SVM speaker verification. In: *Proceedings of The Speaker and Language Recognition Workshop* (Odyssey 2008).

[65] B. Murty, S. R., & Yegnanarayana, B. (2006). Combining evidence from residual phases and MFCC features for speaker recognition. *IEEE signal processing letters, vol. 13*. no.1, 52-55.

[66] Long, Y., Guo, W., and Dai, L. (2008). Interfusing the confused region score of speaker verification systems. *International symposium on Chinese spoken language processing*, 314-317.

[67] Wang, L., Ohtsuka, S., Nakagawa, S. (2009). High improvement of speaker identification and verification by combing MFCC and phase information. *ICASSP*, 4529-4532.

[68] Dehak, N., Kenny, P., Dehak, R., Glembek, O., Dumouchel, P. & Burger, L. (2009). Support vector machines and joint factor analysis for speaker verification. *ICASSP*, 4237-4240.

[69] Gutschoven, B. & Verlinde, P. (2000). Multi-modal identity verification using support vector machines (SVM). In: *Proceeding of ISIF*.

[70] Campbell, W. M., (2002). Genernalized linear discriminant sequence kernels for speaker recognition. *ICASSP*, 161-164.

[71] Campbell, W. M., Sturim, D. E., Reynolds, D. A. & Solomonoff, A. (2006). SVM based speaker verification using a GMM supervector kernel and NAP variability compensation. In *ICASSP*, vol. 1, 97-100.

[72] Yegnanarayana, B., Prasanna, S. R. M., Zachariah, J. M. & Gupta, C. S. (2005). Combining evidence from source, suprasegmental and spectral features for a fixed-text speaker verification system. *IEEE Transaction on speech, and audio processing*, vol.13, no.4, 575-582.

[73] Brummer, N. (2005). Tools for fusion and calibration of automatic speaker detection systems. Available from: http://www.dsp.sun.ac.za/nbrummer/focal/.

[74] NIST 2001. Available from: http://www.itl.nist.gov/iad/mig/tests/spk/.

[75] Cordeiro, H. & Ribeiro, C. (2006). Speaker characterization with MLSF. In: Proceeding of Odyssey, *The Speaker and language recognition workshop*, 1-4.

In: Biometrics: Theory, Applications and Issues
Editor: Ellen R. Nichols, pp. 81-105
ISBN: 978-1-61728-765-7

Chapter 4

FUSION OF LIGHTING INSENSITIVE APPROACHES FOR ILLUMINATION ROBUST FACE RECOGNITION

***Loris Nanni*[1,*] *Sheryl Brahnam*[2,**] *and Alessandra Lumini*[1]**

[1] DEIS, IEIIT—CNR, Università di Bologna, Viale Risorgimento 2, 40136 Bologna, Italy

[2] Computer Information Systems, Missouri State University, 901 S. National, Springfield, MO 65804, USA

Abstract

In this paper the problem of finding a face recognition system that works well both under variable illumination conditions and under strictly controlled acquisition conditions is considered. The problem under consideration has to do with the fact that systems that work well (compared with standard methods) with variable illumination conditions often suffer a drop in performance on images where illumination is strictly controlled. In this chapter we review existing techniques for obtaining illumination robustness and propose a method for handling illumination variance that combines different matchers and preprocessing methods. An extensive evaluation of our system is performed on several datasets (CMU, ORL, Extended YALE-B, and BioLab). Our results show

---

[*] E-mail address: loris.nanni@unibo.it; alessandra.lumini@unibo.it. (Corresponding author)
[**] E-mail address: sbrahnam@missouristate.edu.

that even though some standalone matchers are inconsistent in performance depending on the database, the fusion of different methods performs consistently well across all tested datasets and illumination conditions. Our experiments show that the best result are obtained using gradientfaces as a preprocessing method and orthogonal linear graph embedding as a feature transform.

Keywords: face recognition; illumination conditions; fusion of classifiers.

1. Introduction

Two major challenges in face recognition are illumination and pose variations [38][10]. These are unavoidable sources of noise in many real-world applications where images must be acquired over a span of time and in a variety of circumstances, as is the case, for example, when image acquisition is covert. Illumination effects are the most challenging because they radically alter facial images, as illustrated in Figure 1. In fact, the changes introduced by illumination alter facial images more significantly than the differences manifest in the facial features of different individuals [1][48]. As a result, many researchers are seeking new methods for handling the degradation in performance that illumination variations produce.

It is also the case that many real-world applications acquire images under strictly controlled circumstances, as, for example, in company security systems. Unfortunately, the methods that best handle illumination variations do not perform as well as they could on images acquired under these stricter conditions. In other words, simple methods have not been developed that perform optimally with both types of images.

Face recognition methods that are known to be robust in terms of illumination can be divided into four categories: *heuristic*, *image comparison*, *class-based*, and *3D model-based* [59]. Heuristic approaches to the illumination problem apply simple criteria for dealing with illumination variations. For example, in [9][4], it is shown that discarding the first few principal components results in improved performance when illumination conditions are unstable. Unfortunately, the first few principal components contain important information about the face. As a result, there is a performance drop when the methods are applied to images where the illumination conditions are controlled. Other interesting heuristic approaches are reported in [49], where facial symmetry is used to reconstruct facial textures occluded by dark shadows, and in [15], where a certain number of DCT coefficients are discarded. This method is illumination robust because illumination variations mainly lie in the low-frequency coefficients of the discrete cosine transform (DCT). Unfortunately, in some problems the coefficients with low frequencies offer useful information that is lost [59].

Image-based approaches transform the input face image in a way calculated to reduce the effects of the illumination (for a good survey see [1]). A common approach proposed in the literature is to obtain a representation of the face as the quotient of the image itself with a reference image (see, for example, [24] [29] [44]). In [14][13] an illumination invariant facial feature image for a group of images of the same user is proposed using the total variation based quotient image (TVQI). In [40] a relighting method based on harmonic images is proposed.

A number of image-based approaches proposed in the literature apply various filters to the input image to reduce illumination effects (see, for example, [2][41][17]). Recently, Local Binary Patterns [35][36] have been used to transform the input image. For example, in [47], LBP is coupled with Gabor filters. In [33] a 2D and 3D face recognition system is purposed that is based on a set of AdaBoost feed-forward back-propagation networks trained using LBP. In in [43] the noise robustness of LBP is increased using Local Ternary Patterns (LTP). LTP differs from LBP in that a 3-valued coding (1, 0, -1) is used instead of the binary coding (1,0) of LBP. Other image-based approaches include [34], where a genetic algorithm is used to weigh the most important local classifiers based on Gabor filters.

Class-based approaches are based on *a priori* knowledge of the human face. In [42][22][32][8] a 3D linear illumination subspace for each user was built using three aligned faces acquired under different lighting conditions. In optimal conditions this approach is illumination-invariant. A more effective method is presented in [8][20][19] that is based on the illumination cone. A drawback of methods based on the illumination cone is they require multiple face images of each user. In [6] the face image is segmented into regions, and in each region a specific illumination subspace is calculated. In [21] the authors try to estimate the light-field [28] of the subject's head using an eigenspace technique.

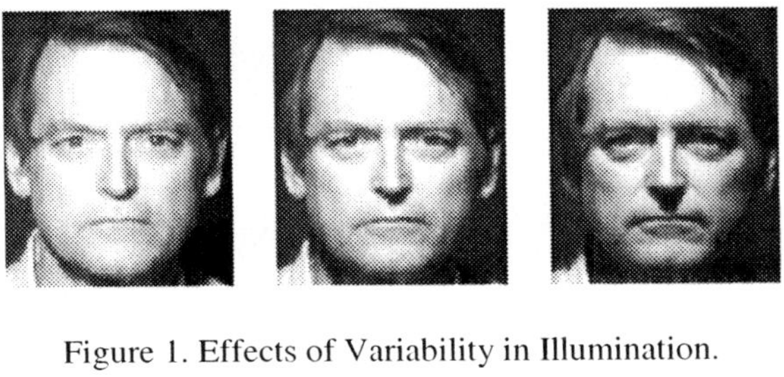

Figure 1. Effects of Variability in Illumination.

Several studies show that the full illumination cone is not needed if more complex techniques based on signal-processing are used in the methods [5][26][27][46]. For example, in [5] it is shown that for a convex Lambertian surface the illumination cone can be accurately approximated by a 9-dimensional linear subspace called the harmonic plane.

The 3D model-based approach uses 3D representations of faces. One of the more widely known 3D approaches is the Shape From Shading (SFS) method, where the three-dimensional shape of a surface is computed from one gray level image of that surface [3][50][21]. A 3D morphable model is proposed in [11], and in [39] the nine spherical harmonic images of the illumination space is recovered from just one image taken under arbitrary illumination conditions.

Methods based directly on 3D scans of faces achieve excellent recognition results since the 3D models provides a very robust representation of the face. However, methods using 3D scans are far more complex compared to methods based on 2D faces.

In this chapter, we focus on the problem of finding simple recognition methods that perform equally well with images acquired under both controlled conditions and under variable illumination conditions. To accomplish our goal we investigate the strengths of combining several preprocessing methods with more than ten different global and local texture descriptors using a multclassifier system. As we will demonstrate, the main benefit of our approach is that, despite its simplicity, we produce results comparable to methods that employ complex 3D models.

The rest of this chapter is organized as follows. In section 2 we present our system architecture, describing our multiclassifier approach and the preprocessing techniques, distance measures, and texture descriptors used in our experiments. In section 3, we describe the face databases (CMU, ORL, Extended YALE-B, and BioLab) that we use in our experiments. In section 4 we present our experimental results as we vary the prepossessing methods and texture descriptors. The robustness of each system is measured using the benchmark datasets. As a result of our experiments we obtain a number of statistically robust observations regarding the effectiveness of our proposed multiclassifier system. Finally, in section 5, we provide a few concluding remarks.

2. Overview of System Architecture

As noted in the introduction, our intention in this chapter is to study the effects that various combinations of feature extractors and preprocessing methods have on a multiclassifier system when using both ideal and ad hoc image acquisition conditions. Our goal is also to find an ensemble of methods that work well with different face recognition problems and benchmark datasets.

Our basic multiclassifier system is outlined in Figure 2. It is divided into four stages: image processing, texture descriptor/feature transformation, distance calculation, and matcher selection and fusion. We defer the description of the preprocessing methods and texture descriptors used in our experiments to sections 2.1 and 22, respectively. In our architecture, local features are extracted using subwindows of dimension 25×25 taken at steps of 11 pixels. In section 2.3 we describe the distant methods used for comparing two templates. As seen in Figure 2, we use Sequential Forward Floating Selection (SFFS) for finding a set of n descriptors that are used to compare the two templates.

SFFS is a bottom up search procedure introduced by [53] consisting of a forward step and a conditional backward step. The forward step starts from an initially empty set of features and successively adds features from a set of original candidates in order to optimize a given objective function. Each time a single feature is added, a backward step is performed that identifies the least significant feature in the current feature set and removes it unless it is the last property added. The number k of retained features is determined according to the objective function, as the minimum number of features that maximizes the performance.

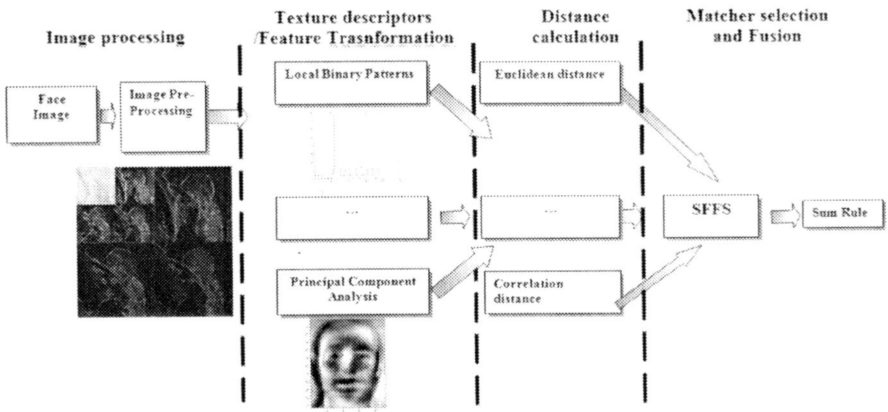

Figure 2. Overview of the proposed system.

Our rationale for employing a multiclassifier system is to exploit the beneficial properties different image-based standalone classifiers possess in order to obtain the highest level of performance and performance stability across our benchmark datasets. The results obtained by our ensemble of matchers are combined using the sum rule.[1]

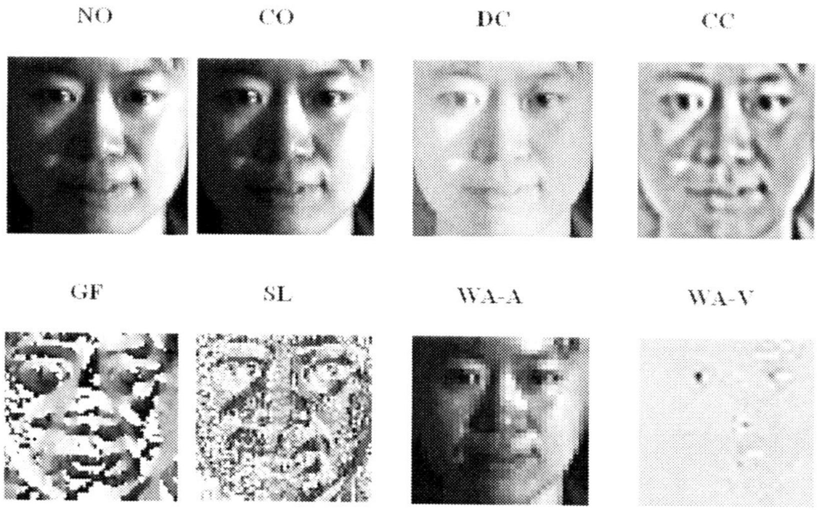

Figure 3. Example images obtained by the different pre-processing.

2.1. Preprocessing Methods

In our experiments, we use the following preprossessing methods: none (NO), Connie (CO), Discrete Cosine Normalization (DC), Contrast correction (CC), GradientFace (GF), Simplified local binary patterns (SL), and Wavelets (WA). This methods are illustrated in Figure 3 and briefly described below.

None (NO) the original image is used without preprocessing for extracting the feature set.

Connie (CO), reported in [16] is an enhancement method, the normalized image is computed by using the operation below:

$$I'(x,y) = \begin{cases} m_n + B \; if \; I(x,y) > m \\ m_n - B \; otherwise \end{cases} \quad B = \sqrt{\frac{n_n (I(x,y) - m)^2}{n}}$$

[1] Before the fusion the scores of each method is normalized to mean 0 and standard deviation 1.

where m and n are mean and variance of the image, m_n and n_n were set to 100.

Discrete Cosine Normalization (DC) is an enhancement method that is based on removing the two AC coefficients with lower frequency, as in [15] where it is shown that that illumination variations can be significantly reduced by truncating low-frequency DCT coefficients in the logarithm DCT domain.

Contrast correction (CC), proposed in [43],[2] is more complex CO and DC. This method is a process of three main steps. In the first step, *gamma correction* is performed. This enhances the local dynamic range of the image in dark or shadowy regions while compressing the image in bright regions and at the highlights. In the second step, the image undergoes a *difference of gaussian filtering*. This is a bandpass filter which reduces aliasing and performs noise contrast equalization. In the third step, *contrast equalization* is performed. This step globally rescales the image intensities to standardize a robust measure of overall contrast or intensity variation.

Gradient Face (GF), proposed in [54],[3] is derived from the image gradient domain. The most interesting property of GF is that fact that it discovers the underlying inherent structure of face images. This is because the gradient domain explicitly considers the relationships between neighboring pixel points.

Simplified local binary patterns (SL), proposed in [55], is a preprocessing method where each pixel is represented with its values in a simplified local binary patterns (the exponential weights are not considered). Since the value that is obtained only accounts for its relative relationship with its neighbors, this makes the result of this method a nonlinear high-pass filter that emphasizes the edges.

Wavelet (WA), is the two-dimensional wavelet transform. In our system we use the biorthogonal wavelet,[4] which is performed by consecutively applying a wavelet transform to the rows and columns of the two-dimensional data. At the first level, the original image is decomposed in four subbands that lead to the following: A, the scaling component containing global low-pass information, and H, V, D, three wavelet components corresponding, respectively, to the horizontal, vertical and diagonal details. In our system, we used the image containing the global low-pass information (WA-A) and the image containing horizontal details (WA-H).

---

[2] The Matlab code is available at http://parnec.nuaa.edu.cn/xtan/Publication.htm
[3] We have used the original matlab code shared by the inventors of the GradientFace
[4] The Matlab 7.0 Wavelet Toolbox has been used in this work.

2.2. Image Descriptors

In our architecture we use both local and global image descriptors. First we describe the local approaches (indentified by a trailing *l* in acronym of the specific methods). Second, we describe the global approaches.

Local Methods Based on Local Binary Patterns (LUl, LRl, LTl) [5]

In the literature, the Local Binary Pattern [36] is an extensively studied descriptor operator. LBP is a histogram that is based on a statistical operator calculated by examining the joint distribution of gray scale values of a circularly symmetric neighbor set of P pixels around a pixel $\mathbf{x}$ on a circle of radius R. In this study we use a multi-resolution descriptor that is obtained by concatenating two histograms calculated with the following parameters: (P=8; R=1) and (P=16; R=2). In our system, we use LBP methods with rotation invariant uniform patterns (*LUl*) and a noise robust version of LBP (*LRl*) where the difference between the gray value of a pixel $\mathbf{x}$ and its neighborhood $\mathbf{u}$ assumes a value of 1 if $\mathbf{u} \geq \mathbf{x} + \tau$, else the difference assumes a value of 0.

A generalization of the Local Binary Patter is the Local Ternary Pattern (*LTl*) [43]. In this method the difference between the gray value of a pixel $\mathbf{x}$ from the gray values in one of its neighborhood $\mathbf{u}$ assumes three values by application of a threshold τ: 1 if $\mathbf{u} \geq \mathbf{x} + \tau$; -1 if $\mathbf{u} \leq \mathbf{x} - \tau$; else 0. The ternary pattern is split into two binary patterns by considering the positive and negative components. The histograms computed from these two patterns are then concatenated for (P=8; R=1) and (P=16; R=2).

In [43][6] *LTl* is not used as LBP for extracting histograms. Rather they show that replacing the local histogram with a local distance transform based similarity metric permits a performance improvement. We name this approach *TAl*.

Local Binary Patterns - Fourier Histogram (LFl)[7]

This method, proposed in [57], it is a rotation invariant image descriptor based on uniform Local Binary Patterns. The discrete Fourier transform is used to extract a class of features that are invariant to the rotation of the input image starting from the histogram rows of the uniform patterns. The descriptors are

[5] For all these LBP methods we varied the LBP Matlab code available at http://www.ee.oulu.fi/mvg/page/lbp_matlab

[6] The Matlab code is available at http://parnec.nuaa.edu.cn/xtan/Publication.htm

[7] The Matlab code is available at http://www.ee.oulu.fi/mvg/page/lbp_matlab

obtained by concatenating two histograms calculated with the following parameters: (P=8; R=1) and (P=16; R=2).

Discrete Cosine Transform (DCL)

This descriptor uses the first 25 discrete cosine transform coefficients to describe the image.

Gabor Filters (GFl)

The 2D Gabor function is a harmonic oscillator, composed of a sinusoidal plane wave of a particular frequency and orientation within a Gaussian envelope. By tuning a Gabor function to a specific frequency and direction, the local frequency and orientation information from an image can be obtained and used as a descriptor [47]

In our experiments the features extracted from the Gabor space are the standard deviations of the image convolved with a bank of 16 Gabor filters. This results in a 16-dimensional feature vector. The filters are obtained considering four scales (σ=1 ,2 ,3 ,4) and, for each scale, four angles ($\theta = 0°, 45°, 90°, 135°$), fixing the frequency to ν=1/3.

Histogram of Gradients (HGl)

This descriptor represents an image using a set of local histograms which count occurrences of gradient orientation in a of the image [62]. We use the weighted version as implemented in [63][8]. In each sub-window the orientation and magnitude of each pixel is calculated. The absolute orientations are discretized over nine equally sized bins in the 0° to 180° range, and the resulting 9-bin histogram is calculated weighting each pixel by the magnitude of its orientation according to the histogram bin.

Maximum Response Filter Banks (MRl)[9]

The filters used In MR*l* are the Gaussian and the Laplacian of Gaussian filters. To achieve rotational invariance, the Maximum Response eight filter bank

---

[8] (he authors have shared the matlab code

[9] The Matlab code is available at http://www.robots.ox.ac.uk/~vgg/research/texclass /code/makeRFSfilters.m

is derived by recording only the maximum filter response across all orientations for the two anisotropic filters [58].

Kernel Laplacian Eigenmaps (KL) [10]

The Laplacianfaces (*LPP*) [23] are the optimal linear approximations to the eigenfunctions of the Laplace Beltrami operator. This descriptor is a global approach preserves local structure and seems to have more discriminating power than the standard principal component analysis (*PC*) approach for classification purposes.

In [18] a kernel version of Laplacianfaces is obtained by simply applying Kernel principal component analysis[11] (KPCA) on the features projected onto the PC space where 98% of the variance is retained. *LP* is then applied on the features transformed by KPCA. In our experiments the kernel PC model with RBF kernel (Delta=2.5) is trained. We use KPCA to project the images onto a lower dimensional space, and then we use LP to project the data.

When we use LP as a standalone feature transform, we reduce the feature space obtained by PC (where 98% of the variance is retained) to a 50-dimensional space. We also use PC as a standalone feature transform, again reducing the original feature space to a reduced space that retains 98% of the variance of the original space.

We also use the orthogonal linear graph embedding (LG). As in the case for LP, we use it to reduce the feature space obtained by PC to a 50-dimensional space.

Neighborhood Preserving Embedding (NP)[10]

This descriptor is a global approach that preserves the local neighborhood structure on the data manifold [56]. Therefore, NP is less sensitive to outliers than PC.

2.3. Distance Measures

As mentioned above, we use several distance measure to compare the distance between two templates.

---

[10] Matlab code available at http://www.cs.uiuc.edu/~dengcai2/Data/data.html

[11] The Statistical Pattern Recognition Toolbox is used in this work, whose Matlab code is publicly available.

Given the vector x_r and x_s the following distances (function *pdist*() of matlab) are tested:

- Euclidean distance (ED)

$$d_{rs}^2 = (x_r - x_s)(x_r - x_s)'$$

- Standardized Euclidean distance (SE)

$$d_{rs}^2 = (x_r - x_s)D^{-1}(x_r - x_s)',$$

where *D* is the diagonal matrix with diagonal elements given by the variance of the variable over the elements.

- Mahalanobis distance (MD)

$$d_{rs}^2 = (x_r - x_s)V^{-1}(x_r - x_s)',$$

where V is the sample covariance matrix.

- City block metric (CB)

$$d_{rs} = \sum_{j=1}^{n} |x_{rj} - x_{sj}|$$

- Cosine distance (CD)

$$d_{rs} = \left(1 - x_r x'_s / (x'_r x_r)^{\frac{1}{2}} (x'_s x_s)^{\frac{1}{2}}\right)$$

- Correlation distance (CO)

$$d_{rs} = 1 - \frac{(x_r - \bar{x}_r)(x_s - \bar{x}_s)'}{[(x_r - \bar{x}_r)(x_r - \bar{x}_r)']^{\frac{1}{2}} [(x_s - \bar{x}_s)(x_s - \bar{x}_s)']^{\frac{1}{2}}}$$

Where:

$$\bar{x}_r = \frac{1}{n}\sum_j x_{rj} \quad \bar{x}_s = \frac{1}{n}\sum_j x_{sj}$$

- Hamming distance (HD)

$$d_{rs} = (\#(x_{rj} \neq x_{sj})/n)$$

- Jaccard distance (JA)

$$d_{rs} = \frac{\#[(x_{rj} \neq x_{sj}) \wedge ((x_{rj} \neq 0) \vee (x_{sj} \neq 0))]}{\#[(x_{rj} \neq 0) \vee (x_{sj} \neq 0)]}$$

- Spearman distance (SP)
  Which is one minus the sample Spearman's rank correlation between observations, treated as sequences of values.
- Chebychev distance (CH)
  Which is simply the maximum coordinate difference.

Moreover, for the GradientFaces and for *TAl*, we use the ad-hoc distances proposed in [54] and [43]. In [43] a complex distance transforms is used for comparing TAl descriptors (see [43] for details). In [54] the similarity between GradientFaces is calculated using the following measure:

$$d_{rs} = \sum_{j=1}^{n} \min(\mathrm{x}_{\mathrm{rj}} - \mathrm{x}_{\mathrm{sj}}, 2\pi - |x_{rj} - x_{sj}|)$$

3. Benchmark Databases

We verify our methods similar to the methods employed in previous work [59] using the ORL Database of Faces [37], YALE-B [7][19][45], EXTENDED YALE-B [61], CMU-PIE [60], and an internal collected dataset (BIOLAB). Below we provide the defining features of each of these databases below, along with the evaluation protocols used in out experiments.

The performance indicator we adopt in our experiments is the Equal Error Rate (EER). ERR is the error rate where the frequency of fraudulent accesses (FAR) and the frequency of rejections that should be correctly verified (FRR) assume the same value. EER is often adopted as a unique measure for characterizing the security level of a biometric system [30].

It should also be noted that the images in the Extended Yale-B ad CMU-PIE databases were cropped and aligned as in http://www.cs.uiuc.edu/~dengcai2/Data/FaceData.html. We did this so that our experiments could easily be compared with others in the literature.

Figure 4. Example images from the *ORL* dataset.

ORL Database

The ORL database contains 10 different images of 40 individuals for a total of 400 images. Example images from the ORL dataset are shown in Figure 4. The facial images were taken at different times and vary in expression, pose, and only slightly in illumination conditions.

In our experiments, results are averaged over five experiments, all conducted using the same parameters. For each experiment we randomly resample the learning and the testing sets (containing respectively half of the patterns). Our protocol using the ORL database deviates from a number of those reported in the literature, such as [59], in that we extract only the face. We did this because we noticed that the background information in the ORL database provides discriminatory information, i.e., oftentimes the background of a given subject was too similar in appearance and we felt this would bias system performance.

YALE-B and Extended Yale-B Database

The YALE-B database contains images of ten individuals under 64 different lighting conditions for nine poses. The updated version, Extended YALE-B, includes images of an additional twenty-eight individuals for a total of 38 subjects. Example from the Extended YALE-B database are shown in Figure 5.

In our experiments, each face image is aligned and cropped to 64×64 pixels. Since our focus is the illumination problem, the only pose used was the frontal face and its varying lighting conditions. The first three images of each user formed the training set and the other images formed testing set.

CMU-PIE Database

The CMU-PIE database contains 41368 images of 68 subjects that vary in pose, illumination, and expression. Example images from the CMU-PIE dataset are shown in Figure 6.

Figure 5. Examples from the *Extended Yale-B* dataset.

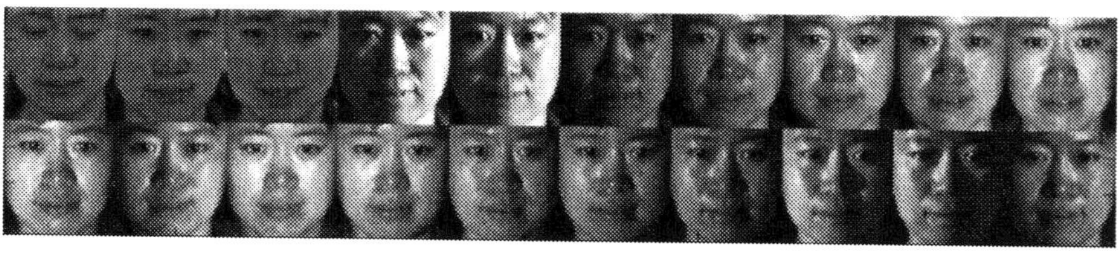

Figure 6. Examples from the *CMU* dataset.

As with the YALE database, all images are cropped and resized to 64×64 and only frontal face images with varying lighting conditions, known as subset C9, are used in our experiments. Subset C9 contains 1428 face images of all 68 subjects, with 24 different illuminations conditions. We use the first three images of each user in the training set and the other images in the testing set.

BioLab Database

The BioLab database was collected in Biometric Systems Lab at the Università di Bologna. It contains 180 images of 10 individuals for a total of 1,800 images. Sixty images per individual were captured in three distinct sessions at least two weeks apart. Example images from the *BioLab* dataset are shown in Figure 6.

In our experiments the face image is automatically located as described in [12]. After cropping, the face-image is resized to 64×64 pixels. The images of individuals captured in the first session formed the training set and the images captured in the other two sessions formed the testing set.

Figure 6. Examples from the *BIOLAB* dataset.

Table 1. The EER obtained by the different pre-processing methods

Pre-processing	NO	CO	DC	CC	GF	SL	WA-A	WA-H
Descriptor	TAI	TAI	NP	PC	LG	LG	TAI	PC
Distance Measure	*	*	CO	SP	ED	ED	*	SP
ORL	8.4	9.2	8.9	17.9	9.7	14.9	8.9	29.4
BioLab	19.4	18.2	12.9	19.5	8.6	17.5	18.5	25.0
Extended Yale-B	35.4	35.1	19.3	5.3	8.3	4.6	35.4	8.9
CMU	13.5	14.9	24.0	16.3	19.3	27.1	12.2	34.0

4. Experimental Results

In the first test, reported in Table 1, we compare the performance of each pre-processing approach, for each method we report the performance obtained by the descriptor and the distance measure that obtain the best average performance in the four datasets using that pre-processing.

In the first test, reported in Table 1, we compare the performance of each descriptor approach, for each method we report the performance obtained by the pre-processing and the distance measure that obtain the best average performance in the four datasets when coupled with that descriptor.

It is very interesting to note that in the ORL dataset (without illumination problem) the best pre-processing is NO (i.e. no one pre-processing is performed). Moreover, it is interesting that in each dataset the best pre-processing is different.

The best average result is obtained by gradientFaces as pre-processing.

In the following Table 3 we compare:

- The best method for each dataset (BEST), in this column each cell contains the EER and the short name of the pre-processing, descriptor and distance measure that are coupled for obtaining the lowest EER in that dataset;
- SA the best average stand-alone method in the four dataset, it is the combination of GF, LG and ED.

A number of experimental findings can be extracted from the results reported in Tables 2-3:

- It is clear the best pre-processing are GF and CC;
- Also in the tests reported in table 2 the best results in each dataset is obtained by a different method.

Now we report the methods tested in our previous work [59] on fusion of matchers for illumination invariant face recognition. In these Tables the following notation is adopted:

- No: no enhancement is performed, classification by EigenFace or by KLEM;
- D1: enhancement based on [15] (removing the 2 AC coefficients with lower frequency), classification by EigenFace or by KLEM;
- Connie: enhancement by the method used in [16], classification by EigenFace or by KLEM;
- Wave: enhancement by the method used in [16], then the two wavelet sub-band are extracted and one EigenFace/KLEM classifier is trained separately on each sub-band. The final recognition is obtained by combining by the sum rule the output of the two EigenFace/KLEM matchers;
- The multi-matcher (MM) obtained by combining the scores of WAVE and D1 according to the sum rule is evaluated. In particular the performance of the multi-matcher is evaluated as a function of the weights assigned to the two classifiers.

Table 2. The EER obtained by the different descriptors

Descriptor	LUl	LTl	LFl	LRl	HGl	DCl	MRl	GFl	TAl	KL	NP	LP	LG	PC
Pre-processing	CC	CC	CC	CC	CC	GF	GF	CC	CC	GF	CC	GF	GF	CC
Distance Measure	CB	SP	CB	SP	CD	ED	CD	SP	*	CO	CD	ED	ED	SP
ORL	8.4	8.2	8.0	9.5	10.5	11.4	16.9	11.9	13.4	14.9	18.9	9.5	9.7	17.9
BioLab	18.2	19.4	16.6	22.1	14.1	18.4	21.2	21.5	18.4	18.5	13.8	9.7	8.6	19.5
Extended Yale-B	38.8	37.8	33.6	38.1	26.5	28.9	23.2	39.5	16.0	9.4	6.1	7.8	8.3	5.3
CMU	34.3	28.4	31.4	31.8	14.0	21.1	23.4	27.5	11.7	19.2	22.2	20.4	19.3	16.3

Table 3. Performance comparison of the new proposed fusion of matchers

Dataset	BEST	SA
ORL	4.97 (WA-A;LTl;CB)	9.7
BioLab	8.6 (GF,LG,ED)	8.6
Extended Yale-B	4.2 (SL,PC,ED)	8.3
CMU	11.7 (CC; TAl;ad-hoc)	19.3

Table 4. The EER on the test sets (EigenFace)

EER	NO	D1	Connie	WAVE	MM	[54]	[43]	NEW	NEW1	NEW2
ORL	7.4	11.4	6.5	5.5	8.5	13.9	13.6	8.5	4.9	7.4
BioLab	16.6	25.8	14.0	16.1	16.1	24.0	17.5	13.2	6.5	10.5
Extended Yale-B	44.4	32.6	42.2	40.7	33.3	31.1	15.7	23.5	4.4	6.9
CMU	46.5	25.7	38.3	38.3	25.5	18.5	11.7	16.7	9.1	11.2

- [54], the method based on gradient faces proposed in [54];
- [43], the method based on the local ternary pattern proposed in [43].
- NEW, the fusion among the set of matchers selected by SFFS using the leave-one-out dataset protocol. The set of descriptors used in a given problem is selected using only the other datasets.
- NEW1, the fusion among the set of matchers selected by SFFS using all the four datasets.
- NEW2, is given by the fusion between [43] and SA.

The most interesting approach is obtained by NEW2, where two only matchers are combined, it dramatically outperforms previous works MM based on KLEM and the state-of-the-arts [43] and [54].

The performance obtained by NEW are not good as those obtained by NEW2, probably this is due to the low number of tested dataset, where each dataset has very different characteristics.

Table 5. The EER on the test sets (KLEM)

EER	NO	D1	Connie	WAVE	MM
ORL	7.1	10.5	6.0	7.4	8.4
BioLab	12.1	20.2	12.9	13.3	14.2
Extended Yale-B	44.0	24.5	39.4	32.8	24.1
CMU	41.1	23.2	39.1	38.5	25.1

Anyway is very interesting the result obtained by NEW2, the reported EER is obtained overfitting the descriptors on the four datasets so the performance is not comparable with the other methods. But we want to stress that exist a set of descriptors that works very well in all the four datasets.

As final test, reported in Table 6, we report for a more wide comparison, the results on the old Yale-B dataset using a standard testing protocol used in several datasets. The face images are divided into four subsets according to the angle between the light source direction and the camera axis. We use Subset 1 as the training set and the other subsets are used for testing. The three subset used as test set are named: SUB-2; SUB-3; SUB-4. In SUB-2 there are weak illumination variations while in SUB-3 and SUB-4 there are strong illumination variations. In this table as performance indicator the error rate is used.

Table 6. Error Rate of state-of-the-art methods on the three test sets from the Yale database

Method	SUB-2	SUB-3	SUB-4
Illumination restoration [29]	0.0	1.7	3.6
9PL [27]	0.0	0.0	2.8
Linear Subspace [19]	0.0	0.0	15.0
Segmented Linear Subspace [6]	0.0	0.0	0.0
Harmonic Images Cast [26]	0.0	0.0	2.7
Illumination cone - Cones Cast [19]	0.0	0.0	0.0
[54]	0.0	0.0	0.0
[43]	0.0	0.0	0.0
NEW2	0.0	0.0	0.0
[54]	0.0	0.0	0.7
[43]	0.0	0.0	0.7
NEW2	0.0	0.0	0.0

Table 7. EER of some local methods when smaller sub-windows are used for extracting features

Dataset	CMU		YALE	
Distance Measure	Euclidean	Cosine	Euclidean	Cosine
SMALL sub-window	15.6	13.1	20.7	16.9
LARGE sub-window	36.1	34.3	39.2	38.8

Please note that the methods whose results are reported in Table 6 (e.g. [19], [26], [6]) require more than one image for each subject in order to allow the model to be calculated. Moreover their performance on databases characterized by "normal" illumination conditions are not known and they are tested using only the old and unreliable Error Rate [30].

As last test we report the performance obtained by *LUI* considering for the local feature extraction small sub-windows of dimension 8×8 taken at steps of 3 pixels. It is clear from the results reported in table 7 that smaller sub-windows dramatically improves the performance. Unfortunately using these small sub-windows it is not feasible, due to the computational power needed, to run all our combinations of pre-processing-texture descriptors.

Finally another problem that should be studied is the performance when only one image for each user is available in the training data as in the FERET datasets, in these datasets the feature transform works poorer than in datasets where more training samples for users are available.

4. Conclusions

In this chapter we have presented an empirical study where different feature extraction approaches are compared and combined. The tested fusion strategy for personal identification using different face matchers permits to obtain performance stability among different test sets (CMU, ORL, Extended YALE-B, BioLab). To obtain this finding we have compared different state-of-the art approaches, several pre-processing methods are coupled with more than ten different texture descriptors then a set of matchers are selected.

The analysis of the state of the art and the experiments carried out in this work suggest that there is not a "best" stand-alone method that performs better than others in all the case studies; we obtain a number of statistically robust observations regarding the robustness of the system, independently on the illumination variations of the dataset, here proposed. Notice that this result is obtained without considering more complex 3D face models.

References

[1] Adini, Y., Moses, Y. & Ullman, S. (1997). "Face Recognition: The Problem of Compensating for Changes in Illumination Direction", *IEEE Transactions On Pattern Analysis and Machine Intelligence*, vol. 19, 721-732.

[2] Arandjelovic, O. & Cipolla, R. (2006). "A New Look at Filtering Techniques for Illumination Invariance in Automatic face recognition", *In proc. IEEE Conference on Automatic Face ad Gesture Recognition*, 449-454.

[3] Atick, J., Griffin, P. & Redlich, N. (1996). "Statistical Approach to Shape from Shading: Reconstruction of Three-Dimensional Face Surfaces from Single Two-dimensional images", *Neural Computation*, vol. 8, 1321-1340.

[4] Bartlett, M. S. & Sejnowski, T. J. (1997). "Independent components of face images: A representation for face recognition", *Proc. of the 4th Annual Jount Symposium on Neural Computation.*

[5] Basri, R. & Jacobs, D. (2001). "Lambertian reflectance and linear subspaces", *In proc. International Conference on Computer Vision*, vol. 2, 383-390.

[6] Batur, A. U. & Hayes, M. H. (2004). III, "Segmented Linear Subspaces for

Illumination-Robust Face Recognition", *International Journal of Computer Vision*, vol. 57, no. 1, 49-66, 2004.

[7] Belhumeur, P. N., Hespanha, J. P. & Kriegman, D. J. (1997). "Eigenfaces vs. Fisherfaces: Recognition Using Class Specific Linear Projection", *IEEE Trans. on Pattern Analysis and Machine Intelligence*, 711-720.

[8] Belhumeur, P. N. & Kriegman, D. J. (1997). "What is the Set of Images of an Object Under All Possible Lighting Conditions?" in Proceedings, IEEE *Conference on Computer Vision and Pattern Recognition*, 52-58.

[9] Belhumeur, P. N., Hespanha, J. P. & Kriegman, D. J. (1997). "Eigenfaces vs. Fisherfaces: Recognition Using Class Specific Linear Projection", *IEEE Trans. on Pattern Analysis and Machine Intelligence*, vol. 19, 711-720.

[10] Blackburn, D. M., Bone, J. M. & Phillips, P. J. (2000). "FRVT 2000 Evaluation Report", 2001, http://www.frvt.org/DLs/FRVT.pdf.

[11] Blanz, V. & Vetter, T. (2003). "Face Recognition Based on Fitting a 3D Morphable Model", *IEEE Transactions on Pattern Analysis and Machine Intelligence, vol. 25*, no. 9, 1-11.

[12] Cappelli, R., Maio, D. & Maltoni, D. (2002). "Subspace classification for face recognition", *Proc. of Workshop on Biometric Authentication - ECCV'02 (BIOW2002)*, 133-141.

[13] Chen, T., Yin, W., Zhou, X., Comaniciu, D. & Huang, T. (2006). "Total variation models for variable lighting face recognition", *IEEE Transactions on Pattern Analysis and Machine Intelligence, vol. 28*, no. 9, 1519-1524.

[14] Chen, T., Yin, W., Zhou, X. S., Comaniciu, D. & Huang, T. S. (2005). "Illumination Normalization for Face Recognition and Uneven Background correction using total variation based image models" *In proc. Proceedings International Conference on Computer Vision and Pattern Recognition*, vol. 2, 532-539.

[15] Chen, W., Er, M. J. & Wu, S. (2006). "Illumination compensation and normalization for robust face recognition using discrete cosine transform in logarithm domain", *IEEE Transactions on Systems, Man and Cybernetics – Part B: Cybernetics*, vol. 36, no. 2, 458-466.

[16] Connie, T., Jin, A. T. B., Ong, M. G. K. & Ling, D. N. C. (2005). "An automated palmprint recognition system", *Image and Vision Computing*, vol. 23, 501-515, 2005.

[17] Du, S. & Ward, R. (2005). "Wavelet-Based Illumination Normalization For Face Recognition", *In proc. International Conference on Image Processing*, vol. 2, 954-957.

[18] Fenga, G., Hua, D., Zhang, D. & Zhou, Z. (2006). "An alternative formulation of Kernel LPP with application to image recognition",

NeuroComputing, vol. 69, 1733-1738.

[19] Georghiades, A., Kriegman, D. & Belhumeur, P. (2001). "From Few to Many: Generative Models for Recognition under Variable Pose and Illumination," *IEEE Trans. Pattern Analysis and Machine Intelligence*, vol. 40, 643-660.

[20] Georghiades, A. S., Kriegman, D. J. & Belhumeur, P. N. (1998). "Illumination Cones for Recognition Under Variable Lighting: Faces", *in Proceedings IEEE Conference on Computer Vision and Pattern Recognition*, 52-58.

[21] Gross, R., Baker, S., Matthews, I. & Kanade, T. (2004). "Face Recognition Across Pose and Illumination", *Handbook of Face Recognition*, In: Z. Stan, Li & K. Anil, Jain, ed., Springer-Verlag.

[22] Hallinan, P. (1994). "A Low-Dimensional Representation of Human Faces for Arbitrary Lighting Conditions", in Proceedings, IEEE *Conference on Computer Vision and Pattern Recognition*, 995-999.

[23] He, X., Yan, S., Hu, Y., Niyogi, P. & Zhang, H. J. (2005). Face Recognition Using Laplacianfaces. IEEE Transactions on Pattern *Analysis and Machine Intelligence*, **27**(3), 328- 340.

[24] Jacobs, D. W., Belhumeur, P. N. & Barsi, R. (1998). "Comparing Images under Variable Illumination", *in Proc. IEEE Conf. On Computer Vision and Pattern Recognition*, 610-617.

[25] Kong, H., Wang, L., Teoh, E. K., Li, X., Wang, J. G. & Venkateswarlu, R. (2005). "Generalized 2D principal component analysis for face image representation and recognition", *Neural Networks*, vol. 18, 585-594.

[26] Lee, K. C., Ho, J. & Kriegman, D. (2005). "Acquiring Linear Subspaces for Face Recognition under Variable Lighting", *IEEE Transactions on Pattern Analysis and Machine Intelligence*, vol. 27, no.5, 684-698.

[27] Lee, K. C., Ho, J., Kriegman, D. (2001). "Nine Points of Lights: Acquiring Subspaces for Face Recognition under Variable Illumination", *In proc. International Conference On Computer Vision and Pattern Recognition*, vol. 1, 519-526.

[28] Levoy, M. & Hanrahan. M. (1996). "Light field rendering", *In proc. Computer Graphics, Annual Conference Series,* 31-41.

[29] Liu, D. H., Shen, L. S., Lam, K. M. & Kong, X. (2005). "Illumination invariant face recognition", *Pattern Recognition*, vol. 38, 1705-1716.

[30] Maio, D., Maltoni, D., Jain, A. K. & Prabhakar, S. (2003). *Handbook of Fingerprint Recognition*, Springer, New York.

[31] Manjunath, B., Chellappa, R. & Von der Malsburg, C. (1992). "A Feature Based Approach to Face Recognition", *In proc. IEEE Conf. On Computer*

Vision and Pattern Recognition, 373-378.

[32] Murase, H., Nayar, S. (1995). "Visual Learning and Recognition of 3D Objects from Appearances", *International Journal of Computer Vision*, Vol. 14, 5-25.

[33] Nanni, L. & Lumini, A. (2007). "Region Boost Learning for 2D+3D based Face Recognition", *Pattern Recognition Letters*, vol.28, no.15, 2063-2070.

[34] Nanni, L. & Maio, D. (2007). "Weighted Sub-Gabor For Face Recognition", *Pattern Recognition Letters*, vol.28, no.4, 487-492.

[35] Ojala, T., Pietikainen, M. & Harwood, D. (1996). "A comparative study of texture measures with classification based on feature distributions", *Pattern Recognition*, vol. 29.

[36] Ojala, T., Pietikainen, M. & Maenpaa, T. (2002). "Multiresolution gray-scale and rotation invariant texture classification with local binary patterns", *IEEE Transactions on Pattern Analysis and Machine Intelligence*, vol. 24, no. 7, 971-987.

[37] ORL database of faces, http://www.uk.research.att.com/ facedatabase.html.

[38] Phillips, P. J., Grother, P., Micheals, R. J., Blackburn, M. D. M. Tabassi, E. & Bone, J. M. (2003). "Face Recognition Vendor Test 2002" Evaluation Report", *NISTIR 6965*, http://www.frvt.org/DLs/FRVT_2002_ Evaluation_Report.pdf.

[39] Qing, L., Shan, S. & Gao, W. (2004). "Eigen-Harmonic Faces: Face Recognition under Generic Lighting", *In proc. International Conference on Automatic Face and Gesture Recognition*, 296- 301.

[40] Qing, L., Shan, S., Gao, W. & Du, B. (2005). "Face Recognition under Generic Illumination based on Harmonic Relighting", *International Journal of Pattern Recognition and Artificial Intelligence*, vol. 19, no. 4, 513-531.

[41] Savvides, M., Kumar, B. V. & Khosla, P. K. (2004). "Corefaces – Robust Shift Invariant PCA based Correlation Filter for Illumination Tolerant Face Recognition", *In proc. International Conference on Computer Vision and Pattern Recognition*, vol. 2, 834-841.

[42] Shashua, A. (1994). "Geometry and Photometry in 3D Visual Recognition", PhD Thesis, MIT.

[43] Tan, X. & Triggs, B. (2007). "Enhanced Local Texture Feature Sets for Face Recognition under Difficult Lighting Conditions", *Proc. IEEE International Workshop on Analysis and Modeling of Faces and Gestures*, 168-182.

[44] Wang, H., Li, S. Z. & Wang, Y. (2004). "Face Recognition under Varying Lighting Conditions Using Self Quotient Image", *In proc. IEEE International Conference on Automatic Face and Gesture Recognition*, pp. 819-824.

[45] Yale Face Database B, http://cvc.yale.edu/projects/yalefaces/ yalefaces.html

[46] Zhang, L. & Samaras, D. (2004). "Pose Invariant Face Recognition Under Arbitrary Unknown Lighting Using Spherical Harmonics", *in Proc. Biometric Authentication Workshop*, 10-23, 2004.

[47] Zhang, W., Shan, S., Chen, X. & Gao, W. (2006). "Are Gabor Phases Really Useless for Face Recognition?", *in proceeding of International Conference on Pattern Recognition*, vol.4, 606-609.

[48] Zhao, W. & Chellappa, R. (1999). "Robust Face Recognition using Symmetric Shape-from-Shading", Technical Report CAR-TR-919, Center for Automation Research, *University of Maryland.*

[49] Zhao, W. (1999). "Improving the Robustness of Face Recognition," *in Proc. 2nd International Conference on Audio- and Video-based Person Authentication*, Washington DC, 78-83.

[50] Zhao, W. (1999). "Robust Image Based 3D Face Recognition", PhD Thesis, *University of Maryland.*

[51] Zhao, W.Y. & Chellappa, R. (2001). "Symmetric Shape-from-Shading Using Self-ratio Image", *International Journal of Computer Vision*, vol. 45, no. 1, 55-75.

[52] Nanni, L. & Lumini, A. (2008). Wavelet Decomposition Tree selection for Palm and Face authentication, *Pattern Recognition Letters*, vol.29, no.3, 343-353, February.

[53] Pudil, P., Novovicova, J. & Kittler, J. (1994). Floating search methods in feature selection. *Pattern Recognition Letters*, **15**, 1119-1125.

[54] Zhang, T., Tang, Y. Y., Fang, B., Shang, Z. & Liu, X. (2009). Face recognition under varying illumination using gradientfaces, *IEEE Trans Image Process*. Nov; **18**(11), 2599-606.

[55] Qian, T. & Veldhuis, R. (2007). Illumination Normalization Based on Simplified Local Binary Patterns for A Face Verification System, *Biometrics Symposium*, 1-6.

[56] Xiaofei He, Deng Cai, Shuicheng Yan, and Hong-Jiang Zhang, "Neighborhood Preserving Embedding", *Tenth IEEE International Conference on Computer Vision* (ICCV'2005).

[57] Ahonen, T., Matas, J., He, C. & Pietikäinen, M. (2009), Rotation invariant image description with local binary pattern histogram fourier features, Image Analysis, SCIA 2009 Proceedings, *Lecture Notes in Computer Science*, **5575**, 61-70.

[58] Geusebroek, J. M., Smeulders, A. W. M. & van de Weijer. J. (2003). Fast Anisotropic Gauss Filtering. IEEE *Transactions on Image Processing*, **12**(8), 938-943.

[59] Annalisa Franco, Loris Nanni, (2009). Fusion of classifiers for illumination robust face recognition, *Expert Systems with Applications*, **36**(5), 8946-8954.

[60] Sim, T., Baker, S. & Bsat, M. (2002). "The CMU pose, illumination, and expression (PIE) database," in Proc. IEEE Int. Conf. *Automatic Face and Gesture Recognition*, 46-51.

[61] Lee, K., Ho, J., Kriegman, D. (2005). Acquiring linear subspaces for face recognition under variable lighting. IEEE TPAMI, **27**(5), 684-698.

[62] Dalal, N. & Triggs, B. (2005). Histograms of oriented gradients for human detection, in: *Proceedings of the 9th European Conference on Computer Vision*, San Diego, CA.

[63] Poppe. R. (2007). Evaluating Example-based Pose Estimation: Experiments on the Human Eva Sets, CVPR 2nd Workshop on *Evaluation of Articulated Human Motion and Pose Estimation*.

In: Biometrics: Theory, Applications and Issues ISBN: 978-1-61728-765-7
Editor: Ellen R. Nichols, pp. 107-122

Chapter 5

A Two-Part Generalized Linear Mixed Modelling Approach to Analyze Physical Activity Outcomes

***Andy H. Lee*[1,*]*, Liming Xiang*[2] *and Fumi Hirayama*[1]
[1] School of Public Health, Curtin Health Innovation Research Institute, Curtin University of Technology, Perth, WA, Australia
[2] Division of Mathematical Sciences, SPMS, Nanyang Technological University, Singapore

Abstract

Physical activity (PA) is a modifiable lifestyle factor for many chronic diseases and its health benefits are well known. PA outcomes are often measured and assessed in many clinical and epidemiological studies. This chapter first reviews the problems and issues regarding the analysis of PA outcomes. These include outliers, presence of many zeros and correlated observations, which violate the statistical assumptions and render standard regression analysis inappropriate. An alternative two-part generalized linear mixed models (GLMM) approach is proposed to analyze the heterogeneous and correlated PA data. At the first part, a logistic mixed regression model is fitted to estimate the prevalence of PA and factors associated with PA participation. A gamma mixed regression model is adopted at the second part to assess the effects of predictor variables among those

* E-mail address: Andy.Lee@curtin.edu.au, Tel: +61 8 92664180, Fax: +61 8 92662958, Address: School of Public Health, Curtin University of Technology, GPO Box U 1987, Perth, WA, 6845, Australia. (Corresponding author)

with positive PA outcomes. Variations between clusters are accommodated by random effects within the GLMM framework. The proposed methods are demonstrated using data collected from a community-based study of elderly PA in Japan. The findings provide useful information for targeting physically inactive population subgroups for health promotion programs.

1. Introduction

The importance and health benefits of physical activity are well known. PA is a modifiable lifestyle factor for many chronic diseases including stroke, cardiovascular diseases, osteoporosis and various cancers. It is estimated that half of all functional decline associated with the ageing process is preventable if adequate levels of PA are maintained (O'Brien Cousins, 2003), confirming the contribution of PA to postponing disability and improving survival. PA is also known to enhance psychological and emotional well-being (Adams, Bezner & Whistler, 1999), as well as increasing satisfaction with life (Rejeski & Mihalko, 2001). In relation to mental health, evidence has suggested that PA can improve quality of life by lowering levels of psychological distress and negative moods (Kritz-Silverstein, Barrett-Connor & Corbeau, 2001). On the other hand, a sedentary lifestyle or physical inactivity is responsible for 1.9 million deaths globally each year according to the World Health Organization (2003). The Centers for Disease Control in the USA have developed guidelines on PA for adults, recommending a minimum of 30 minutes of moderate activity daily (Pate et al, 1995). Similar recommendations have been proposed in Australia (Department of Health and Ageing, 2009) and other countries to promote regular PA.

This book chapter first reviews methodological issues concerning the analysis of PA outcomes in Section 2. These include outliers and non-normality, presence of many zero observations, and violation of the independence assumption, so that application of the standard regression model may lead to spurious associations and misleading conclusions. In Section 3, a two-part generalized linear mixed modeling approach is proposed to analyze the heterogeneous and correlated PA data. Motivated by our empirical study of PA in daily life among older adults, the methodology is illustrated using an actual data set in Section 4. Further discussion of modelling issues and extensions of the method are given in Section 5.

2. Physical Activity Assessment

Habitual PA levels are often assessed and monitored in many clinical, intervention and observational studies, in view of the importance of PA to human health and well being. Measurement of PA comprises duration, frequency and intensity. The data are either collected by means of self-report using a validated questionnaire, or objectively measured from accelerometry (Ekelund et al, 2006). Typically, PA outcomes are expressed in terms of time (e.g. minutes per week) or energy expenditure as metabolic equivalent tasks (MET). The MET score is the ratio of metabolic rate during the activity compared with the metabolic rate during rest. Each type of activity (vigorous, moderate, walking, etc) is allocated a MET score under a standard compendium (Ainsworth et al, 2000). MET-minutes are computed by multiplying the MET score by the minutes performed at each activity, then multiplied by the frequency of the activity to obtain MET-min/week.

In many biomedical studies, the primary interest is to model PA outcomes and correlate PA with demographic, social, environmental or clinical factors. Regression-type analyses are often applied. However, there are several methodological issues concerning their application.

2.1. Outliers and Non-Normality

The distribution of the continuous PA variable is highly skewed (Rzewnicki et al, 2003; Jurj et al, 2007), with the presence of extreme scores and outliers, severely violating the normality assumption (Harrison, Hayes & Newman, 2009). Therefore, sample means must be presented with caution. Indeed, given the non-normal distribution of energy expenditure in many populations, it has been suggested that the continuous PA variable be reported in medians and interquartile ranges rather than means (IPAQ, 2005). Median and quartile values are presented in some recent studies (Harrison, Hayes & Newman, 2009; Allman-Farinelli et al, 2009).

The standard approach to deal with skewness of the PA distribution is to remove outliers and apply some transformation such as logarithm before performing linear regression (Ekelund et al, 2006; Rzewnicki et al, 2003). Unfortunately, trimming rule to discriminate outlying observations and the selection of appropriate transformation are often determined arbitrarily without theoretical justification or support (IPAQ, 2005). Alternatively, observed PA levels are dichotomized at the median to create binary outcomes for logistic regression analysis (Jurj et al, 2007; Nitzan Kaluski et al, 2009). Likewise, PA levels are sometimes categorized into three groups: zero or inactive, insufficient (less than a specified recommended level), sufficient (achieving the recommended level) prior to applying multinomial logistic regression (Shibata et al, 2009). It should be remarked, however, that the conversion of continuous PA into discrete random variables results in loss of accurate information about these outcomes.

2.2. High Prevalence of Zeros

At least 60% of the world's population fails to achieve the minimum recommendation of 30 minutes of moderate-intensity PA daily, according to the World Health Organization (2004). In particular, elderly people are less likely to be active, believing that the benefits associated with PA are outweighed by the hardships or barriers, or that the environment does not offer attractive PA options. In the USA, 51% of adults aged over 65 years are inactive (US Department of Health and Human Services, 2001). Similarly, 46% of Australian adults aged 60 to 75 years are found to be inactive and about one-third of them are completely sedentary (Bauman et al, 2002). This holds true especially for those with medical conditions or suffering from illness and chronic disease(s) (Harrison, Hayes & Newman, 2009; Hirayama et al, 2008). Moreover, scientific evidence has indicated that episodes or bouts of at least 10 minutes of PA are required to achieve health benefits, so that responses of less than 10 minutes are often re-coded to "zero" (IPAQ, 2005). The high prevalence of sedentary subjects present in the sample will lead to a preponderance of zeros for PA in the data set, which in turn will pose difficulties for modelling and analysis using linear regression.

2.3. Lack of Independence

For subjects residing in the same district or catchment area, correlations between their PA levels are expected due to similar socio-economic status and environmental conditions. In longitudinal studies, repeated PA measures are recorded for individuals in the cohort; these (within subject) observations are also likely to exhibit high correlation. Under these settings when the independence assumption is violated, application of standard regression analysis will lead to biased parameter and standard error estimates, resulting in spurious associations and possible misleading inferences concerning the PA outcome, for example, in terms of identifying pertinent risk factors or evaluating the effectiveness of preventive interventions. Adjustment for clustering effects to account for the inherent data dependency is deemed necessary (Allman-Farinelli et al, 2009).

3. Two-Part Generalized Linear Mixed Models

In many studies, the objectives are: (1) to estimate the prevalence of PA and/or factors affecting PA participation; (2) to assess the effects of pertinent variables on PA level among subjects with positive PA exposure. A two-part generalized linear mixed modeling (GLMM) approach is constructed below to achieve both purposes.

Suppose that a subject engages in a certain level of PA with probability π. Let binary variable Z_{ij} = 1 for individual j within cluster i whose PA > 0, and Z_{ij} = 0 for individuals who are completely sedentary. Let $\pi_{ij} = \pi(x_{ij}, \alpha) = P(Z_{ij} = 1 \mid x_{ij})$ for given covariates x_{ij} and associated parameters α. Sedentary individuals contribute a likelihood factor of $1 - \pi_{ij}$, whereas those who engage in PA contribute a likelihood factor $\pi_{ij} f(y_{ij}, w_{ij}, \beta)$, where f represents the probability density of the non-zero PA levels Y_{ij} ($j = 1,2,\ldots,n_i$, $i = 1,2,\ldots,m$) with covariates w_{ij} and parameters β. The likelihood function $L(\alpha, \beta)$ factorizes into a probability component which concerns only the aspect of participation,

$$L_1(\alpha) = \prod_{Z_{ij}=0} (1 - \pi_{ij}) \prod_{Z_{ij}=1} \pi_{ij},$$

and a conditional component which concerns realization of the PA event,

$$L_2(\beta) = \prod_{Z_{ij}=1} f(y_{ij}, w_{ij}, \beta).$$

Hence, the above objectives can be treated completely separately. Such a two-part conditional approach is in principle analogous to the hurdle model used for modeling discrete counts with extra zeros (Welsh et al, 1996; Yau & Lee, 2001; Lee et al, 2003; Yau, Wang & Lee, 2003).

A natural way to relate π with covariate vector x is via the logistic regression model:

$$\log(\pi_{ij} / (1 - \pi_{ij})) = x_{ij}' \alpha,$$

whereas the gamma distribution naturally accommodates different degrees of skewness of the positive PA variable Y_{ij} through its scale parameter θ.

$$f(y_{ij}) = \frac{1}{\Gamma(\theta)}\left(\frac{\theta}{\mu_{ij}}\right)^{\nu} y_{ij}^{\nu-1} \exp\left(-\frac{y_{ij}\theta}{\mu_{ij}}\right),$$

and μ_{ij} is the mean of Y_{ij}. The variance of Y_{ij} is given by $Var(Y_{ij}) = \phi\mu_{ij}^2$ with $\phi = \theta^{-1}$. The gamma regression model relates μ_{ij} to the covariates w_{ij} by a link function:

$$g(\mu_{ij}) = w'_{ij}\beta ,$$

and β is the corresponding unknown coefficient vector. Alternatively, a log-normal regression model can be formulated by adopting a normal distribution for the log-transformed Y_{ij}.

It is anticipated that the observations exhibit some within cluster correlation. Generalized linear mixed models (McGilchrist, 1994), which extend the linear regression model, are useful for modeling such clustered data. The GLMM approach accommodates and adjusts for the inherent correlation of the observations, thus avoids misleading conclusions being drawn due to violation of the independence assumption. The correlation structure may be modeled explicitly by random effects u and v attached to each cluster, viz.

$$\log(\pi_{ij}/(1-\pi_{ij})) = x'_{ij}\alpha + u_i ,$$

$$g(\mu_{ij}) = w'_{ij}\beta + v_i .$$

For simplicity of estimation, u and v may be assumed to be independent and normally distributed with variance σ_u^2 and σ_v^2, respectively. The likelihood $L(\alpha,\beta)$ can be factorized into two parts, one being the logistic and the other corresponds to the conditional probability density function f (e.g. gamma). Following the method of zero-augmented gamma mixed model for longitudinal data with many zeros (Yau, Lee & Ng, 2002), for conditionally fixed u and v, the two parts then reduce to two separate GLMM applications. In particular, the best linear unbiased prediction (BLUP) type log-likelihood of the logistic part is given by $l_1 + l_u$, where

$$l_1 = \log L_1(\alpha \mid u),$$

$$l_u = -\tfrac{1}{2}\left(n \log(2\pi\sigma_u^2) + \sigma_u^{-2} u'u\right).$$

Similarly, the BLUP type log-likelihood of the conditional gamma component is given by $l_2 + l_v$, where

$$l_2 = \log L_2(\beta \mid v),$$

$$l_v = -\tfrac{1}{2}\left(n \log(2\pi\sigma_v^2) + \sigma_v^{-2} v'v\right).$$

Since both parts belong to the exponential family of distributions, with the orthogonal parameterization, it is then fully efficient to fit the components separately. Residual maximum quasi-likelihood estimates of the parameters can be obtained by fitting the data separately within the GLMM framework (Yau, Lee & Ng, 2002). Therefore, model fitting for the two parts can be readily implemented in standard statistical packages, such as SAS by invoking procedure GLIMMIX. The corresponding SAS instructions are given in the Appendix.

4. Example

It is important to determine the level of PA and its correlates in order to target at-risk population subgroups and to develop tailored interventions (Nitzan Kaluski et al, 2009; Shibata et al, 2009). A community-based study was conducted to investigate PA in daily life among Japanese older adults (Hirayama, Lee & Binns, 2008). A total of 575 eligible participants (355 men and 220 women) aged 55 to 75 years were recruited from 10 prefectures in 2006. Information on PA levels was obtained by face-to-face interviews using the International Physical Activity Questionnaire (IPAQ, 2005). Total PA was defined as the sum of walking, moderate and vigorous activities which last more than 10 minutes. Overall, 114

(20%) of these community-dwelling subjects were inactive and did not participate in regular exercise, i.e. PA = 0. The median (total) PA was 693 (interquartile range 1308) MET min/week, which was below the Japanese government recommendation of 1380 MET min/week for individuals (Ministry of Health, Labour and Welfare, 2006). As shown in Figure 1, the distribution of PA was extremely skewed for those participants whose observed PA > 0 ($n = 461$).

The available covariates for this study were body mass index (in kg/m^2), gender (0 = male, 1 = female), age group (55-59, 60-64, 65-69, 70-75 years), education (0 = high school or below, 1 = college/university), marital status (0 = single/divorced, 1 = married), employment status (0 = unemployed, 1 = employed), smoking (0 = non-smoker, 1 = current smoker), and presence of comorbidity (0 = no, 1 = yes). In this sample, the majority of participants were men (62%), the mean age was 63.2 (SD 6.4) years and mean BMI 23.3 (SD 3.1). Most of them (79%) were married and had attained high school education (62%). The prevalence of smoking was 18%. About 38% of them were not working, and half (49%) had a health condition such as hypertension, ischemic stroke, diabetes mellitus, depression or cancer.

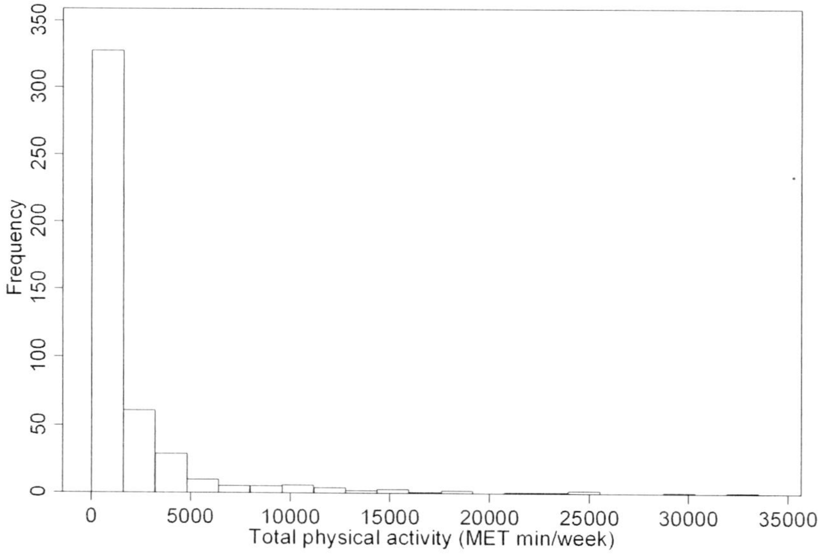

Figure 1. Empirical distribution of PA level for the subgroup of participants who engaged in regular PA ($n = 461$).

Table 1. Results from fitting (a) gamma mixed regression model and (b) linear regression model to PA for Japanese older adults who engaged in regular PA (n = 461)

	(a) Gamma mixed regression with log link			(b) Linear regression with log-transformation		
Variable	Coefficient	Standard error	p value	Coefficient	Standard error	p value
Intercept	8.141	0.662	<0.001	7.255	0.542	<0.001
Body mass index	-0.029	0.023	0.202	-0.021	0.020	0.282
Gender: female	-0.100	0.170	0.556	0.010	0.144	0.945
Marital status: married	-0.036	0.189	0.850	0.002	0.163	0.988
Smoking: current smoker	0.006	0.197	0.976	0.035	0.170	0.834
Co-morbidity: yes	-0.060	0.143	0.673	-0.184	0.123	0.134
Education: college/university	-0.444	0.154	0.004	-0.187	0.133	0.159
Employment status: employed	0.342	0.172	0.047	0.017	0.147	0.910
Age group: 60-64	0.602	0.197	0.002	0.452	0.169	0.008
65-69	0.390	0.227	0.087	0.402	0.191	0.036
≥ 70	0.085	0.228	0.708	0.127	0.195	0.517

The logistic mixed regression model was first fitted to the binary PA data Z_{ij} clustered within prefectures i = 1,..., 10. None of the covariates were found to be significant affecting the probability π. The resulting constant linear predictor model gave $\hat{\alpha}$ = 2.754 with standard error 0.942 and random effect variance estimate $\hat{\sigma}_u^2$ = 4.635. The gamma mixed regression model was next fitted to the PA outcomes $Y_{ij} > 0$ via a log-link function $g(\mu_{ij}) = \log(\mu_{ij})$. Table 1(a) presents the regression results. The estimate of scale parameter θ is 2.158 and $\hat{\sigma}_v^2$ = 0.243. Education level, employment status and age group were significant factors associated with PA. Specifically, those aged 60-64 years had significantly higher PA levels than the reference group aged 55-59 years, whereas unemployed adults and participants who were college/university graduates tended to have relatively lower PA levels. Similar results were obtained when an inverse link function $g(\mu_{ij}) = 1/\mu_{ij}$ was used instead of the log-link. However, as shown in Table 1(b), education level and employment status became non-significant under the standard linear regression model with log-transformed PA. This suggested that logarithmic transform alone was insufficient for handling such highly skewed and heterogeneous PA data. Finally, to assess the adequacy of the fitted gamma mixed regression model, the corresponding scaled Pearson residuals are plotted against the fitted values in Figure 2, which indicates that no systematic departure is evident.

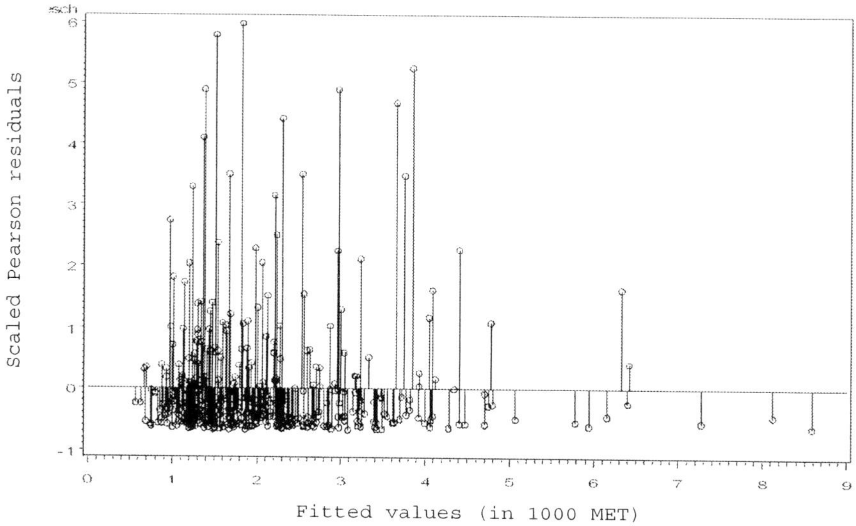

Figure 2. Plot of scaled Pearson residuals against fitted values from the gamma mixed regression model for the subgroup of participants who engaged in regular PA (n = 461).

5. Discussion

Critical assessment of PA outcomes is fundamentally important in ensuring that the correct conclusion is made and appropriate action being implemented. It is at an analyst's peril that regression methods are applied without due consideration of the underlying issues and assumptions. The standard practice is to apply linear or mixed regression to the log-transformed PA outcomes, though such analysis is sensitive to extreme outliers and the presence of many zeros. However, the literature has provided little guidance to deal with the various methodological issues. This book chapter is concerned with the appropriate assessment of PA. As evident from the empirical study, the proposed method will enable healthcare professionals to focus on substantive issues and to draw accurate and valid conclusions inferred from the correlated and heterogeneous PA outcomes. In the presence of anomalous observations, the two-part GLMM method can provide insights into the sources of heterogeneity and excess zeros, while accommodating the inherent correlation due to the clustering data structure. In addition, the detection of sedentary individuals or subgroups will provide further insights into the nature of the problem.

Many factors influence PA behaviour. Successful development and implementation of interventions to increase PA levels can benefit from understanding the associated factors (Jancey et al, 2008; Shibata et al, 2009). Although our empirical study did not comprehensively evaluate demographic and lifestyle factors, additional population subgroups have been highlighted based on the gamma mixed regression analysis. By highlighting population subgroups for attention, useful information can be obtained for targeting physically inactive people in terms of health promotion programs. Appropriate PA guidelines can then be developed specifically for older adults in order to maintain their health and well being (Hirayama, Lee & Binns, 2008).

In the second part of the GLMM model, an inverse Gaussian distribution can be assumed as an alternative to the gamma distribution to deal with the highly skewed PA observations. Logistic-inverse Gaussian regressions can be defined in a similar manner. The proposed two-part GLMM approach provides a practical and statistically valid method to analyze heterogeneous and clustered PA data. Fitting of this class of models can be readily performed using the SAS package. The methodology can be extended to handle repeated measures of PA and accommodate other hierarchical or multi-level settings by appropriate specification of the random components. These and other generalizations will be pursed in future research.

Acknowledgment

This research is supported by a project grant from the National Health and Medical Research Council of Australia.

Appendix

SAS Instructions for Fitting Two-Part GLMM

PROC GLIMMIX is invoked to fit a logistic mixed regression model for binary outcome *bipa* in part 1 and a gamma mixed regression model for continuous response *totalmet* in part 2.

```
data PA0;
set PA;
bipa=totalmet;
if totalmet=0 then bipa=0;
else if totalmet NE 0 then bipa=1;
catage=age;
if age<55 then catage=.;
else if age>=55 and age<=59 then catage=4;
else if age>=60 and age<=64 then catage=3;
else if age>=65 and age<=69 then catage=2;
else catage=1;
...
...
/*1) Fitting logistic regression with random effects
*/
%glimmix(data=PA0,
        procopt=method=reml,
        stmts=%str(
        class catage prefecture;
        model bipa=gender catage bmi education
      employment marital smoking
      comorbidity/solution;
        random prefecture/solution;
        ),
        error=binomial,
        link=logit,
        out=logitglmmout,
        maxit=100
);
```

```
run;

/*2) Fitting gamma regression with random effects */
data gpa;
set pa0;
where (bipa NE 0);
run;
%glimmix(data=gpa,
          procopt=method=reml,
          stmts=%str(
          class catage prefecture;
           model totalmet=gender catage bmi education
        employment          marital          smoking
        comorbidity/solution;
          random prefecture/solution;
          ),
        error=gamma,
           link=log,
           out=gglmmout,
           maxit=100
);
run;
```

References

Adams, T. B., Bezner, J. R. & Whistler, L. S. (1999). The relationship between physical activity and indicators of perceived wellness. *American Journal of Health Studies*, **15**, 130-138.

Ainsworth, B. E., Haskell, W. L. & Whitt, M. C. et al. (2000). Compendium of physical activities: an update of activity codes and MET intensities. *Medicine & Science in Sports & Exercise*, **32**, S498-S504.

Allman-Farinelli, M. A., Chey, T. & Merom, D. et al. (2009). The effects of age, birth cohort and survey period on leisure-time physical activity by Australian adults: 1990-2005. *British Journal of Nutrition*, **101**, 609-617.

Bauman, A., Bellew, B. & Vita, P. et al. (2002). Getting Australia active: Towards better practice for the promotion of physical activity. http://www.dlg.nsw.gov.au/Files/Information/Getting%20Australia%20Active.pdf.

Department of Health and Ageing. (2009). *Physical Activity Guidelines*. Commonwealth of Australia, Canberra.

Ekelund, U., Sepp, H. & Brage, S. et al. (2006). Criterion-related validity of the last 7-day, short form of the International Physical Activity Questionnaire in Swedish adults. *Public Health Nutrition*, **9**, 258-265.

Harrison, S., Hayes, S. C. & Newman, B. (2009). Level of physical activity and characteristics associated with change following breast cancer diagnosis and treatment. *Psychooncology*, **18**, 387-394.

Hirayama, F., Lee, A. H. & Binns, C. W. et al. (2008). Physical activity of patients with chronic obstructive pulmonary disease: implications for pulmonary rehabilitation. *Journal of Cardiopulmonary Rehabilitation and Prevention*, **28**, 330-334.

Hirayama, F., Lee, A. H. & Binns, C. W. (2008). Physical activity of adults aged 55 to 75 years in Japan. *Journal of Physical Therapy Science*, **20**, 217-220.

International Physical Activity Questionnaire (IPAQ). (2005). Guidelines for data processing and analysis of the international physical activity questionnaire (IPAQ). http://www.ipaq.ki.se/ipaq.htm.

Jancey, J. M., Lee, A. H. & Howat, P. A. et al. (2008). The effectiveness of a physical activity intervention for seniors. *American Journal of Health Promotion*, **22**, 318-321.

Jurj, A. L., Wen, W. & Gao, Y. T. et al. (2007). Patterns and correlates of physical activity: a cross-sectional study in urban Chinese women. *BMC Public Health*, **7**, 213.

Kritz-Silverstein, D., Barrett-Connor, E. & Corbeau, C. (2001). Cross-sectional and prospective study of exercise and depressed mood in the elderly: the Rancho Bernardo study. *American Journal of Epidemiology*, **153**, 596-603.

Lee, A. H., Wang, K. & Yau, K. K. W. et al. (2003). Truncated negative binomial mixed regression modelling of ischaemic stroke hospitalizations. *Statistics in Medicine*, **22**, 1129-1139.

McGilchrist, C. A. (1994). Estimation in generalised mixed models. *Journal of the Royal Statistical Society B*, **56**, 61-69.

Ministry of Health, Labour and Welfare. (2006). *Standard of physical activity for healthy life 2006*. http://www.mhlw.go.jp/shingi/2006/07/dl/s0725-9e.pdf.

Nitzan Kaluski, D., Demem Mazengia, G. & Shimony, T. et al. (2009). Prevalence and determinants of physical activity and lifestyle in relation to obesity among schoolchildren in Israel. *Public Health Nutrition*, **12**, 774-782.

O'Brien Cousins, S. (2003). Grounding theory in self-referent thinking: conceptualizing motivation for older adult physical activity. *Psychology of Sport and Exercise*, **4**, 81-100.

Pate, R. R., Pratt, M. & Blair, S. N. et al. (1995). Physical activity and public health. A recommendation from the Centers for Disease Control and Prevention and the American College of Sports Medicine. *JAMA*, **273**, 402-407.

Rejeski, W. J. & Mihalko, S. L. (2001). Physical activity and quality of life in older adults. *Journal of Gerontology A – Biological Sciences and Medical Sciences*, **56**, 23-35.

Rzewnicki, R., Vanden Auweele, Y. & De Bourdeaudhuij, I. (2003). Addressing overreporting on the International Physical Activity Questionnaire (IPAQ) telephone survey with a population sample. *Public Health Nutrition*, **6**, 299-305.

Shibata, A., Oka, K., Nakamura, Y. & Muraoka, I. (2009). Prevalence and demographic correlates of meeting the physical activity recommendation among Japanese adults. *Journal of Physical Activity and Health*, **6**, 24-32.

US Department of Health and Human Services. (2001). *Healthy people 2010-Reproductive Health*. http://www.hhs.gov/opa/pubs/hp2010_rh.html.

Welsh, A. H., Cunningham, R. B. & Donnelly, C. F. et al. (1996). Modelling the abundance of rare species: statistical models for counts with extra zeros. *Ecological Modelling*, **88**, 297-308.

World Health Organization. (2003). Annual Global Move for Health Initiative: A concept paper. www.who.int/moveforhealth/publications/en/mfh_concept_paper_english.pdf.

World Health Organization. (2004). *Benefits of Physical Activity*. http://www.who.int/dietphysicalactivity/factsheet_benefits/en/.

Yau, K. K. W. & Lee, A. H. (2001). Zero-inflated Poisson regression with random effects to evaluate an occupational injury prevention programme. *Statistics in Medicine*, **20**, 2907-2920.

Yau, K. K. W., Lee, A. H. & Ng, A. S. K. (2002). A zero-augmented gamma mixed model to analyse longitudinal data with many zeros. *Australia & New Zealand Journal of Statistics*, **44**, 177-183.

Yau, K. K. W., Wang, K. & Lee, A. H. (2003). Zero-inflated negative binomial mixed regression modeling of over-dispersed count data with extra zeros. *Biometrical Journal*, **45**, 437-452.

In: Biometrics: Theory, Applications and Issues ISBN: 978-1-61728-765-7
Editor: Ellen R. Nichols, pp. 123-141

Chapter 6

BIOMETRIC IDENTIFICATION PARADIGM: TOWARDS PRIVACY AND CONFIDENTIALITY PROTECTION*

Julien Bringer[1,†] ***and Hervé Chabanne***[1,2,‡]
[1] Morpho
[2] Télécom ParisTech

Abstract

In the last years, numerous research papers have considered the problem of privacy protection in biometric authentication systems (1-to-1). When someone wants to authenticate himself, the fresh biometric data he presents has to match with the reference related to the one he claims to be. Many proposals have thus been made to ensure the confidentiality of the biometric data involved during this verification.

Biometric identification enables to identify an individual among many others without requiring any prior claim of identity (1-to-many). Typically, it consists in checking the belonging of a user to a database. Paradoxically, today, identification constitutes the main application of biometric systems whereas privacy protection in such identification systems has received relatively little attention in the literature. In this chapter, we show

---

*This work is partially supported by funding under the Seventh Research Framework Programme of the European Union, Project TURBINE (ICT-2007-216339).

†E-mail address: julien.bringer@morpho.com

‡E-mail address: herve.chabanne}@morpho.com

how to extend the previous works on biometric authentication in order to also cover biometric identification while maintaining the privacy of the individuals. This is not a trivial task as we want to go much faster than performing comparisons with all the biometric references in the system, i.e. an exhaustive search. The main difficulty comes here from the fact that as we must deal with data that are protected, we cannot merely rely on traditional biometric identification methods. This can be overcome successfully thanks to new techniques that have been suggested recently in a few papers.

Keywords: Identification, Biometrics, Privacy, Searchable Encryption.

1. Introduction

Maintaining privacy of users during their biometric authentication has received a lot of attention during the last years. As biometric data can serve to track people among the different applications that use them, methods to ensure their confidentiality are needed. Data can be encrypted to reach this end but this opens the issue of their comparison after this encryption. A response to this problem has been given by the introduction of secure sketches [27, 20]. However [8, 35] point out that the security of secure sketches is not well etablished in practice. [9, 14, 15, 36, 7] then suggest to combine secure sketches and homomorphic encryption. More precisely, the simplicity of the comparison of biometric data induced by secure sketches (often a simple algebraic operation) is directly transformed by the homomorphic property of the cryptosystem into an operation to be performed on the ciphertexts.

To authenticate themselves, users have to log in to the system which checks that they are who they are pretending to be. While this way of proceeding is quite ordinary, it constitutes only one part of what is possible when dealing with biometric recognition. Biometric identification is another common practice. In this case, the user has only to provide a new capture of his biometric trait to identify himself. For people working on biometric systems, identification may seem more natural. For instance, the convenience of having nothing to remember or possess is a nice feature. However, until recently, they are very few attempts [1, 12, 13] in order to generalize existing authentication techniques to the identification context.

The goal of this chapter is to make a synthesis of what has been proposed so far. It is organized as follows.

We begin by recalling some basic facts about biometric identification and introducing some notations in Section 2. We here focus on the description of existing techniques working on biometric data in clear. We then define the security properties we want to reach in Section 3. Section 4. goes deeper into details and gives the technical tools needed by our protocols. Section 5. describes some existing protocols which fulfill our goals. Section 6. concludes by elaborating on the issues which are still open.

2. Biometric Identification with Quantized Biometric Data in Clear

Biometric data are captured by biometric sensors. We write $b \leftarrow \beta$ to indicate that when we are capturing the biometric trait β of an user, we obtain the value b. As these captures are sensitive to variations, when a second capture of the same biometric trait β is made, we obtain a different value $b' \leftarrow \beta$, $b' \neq b$.

In the following, for a given biometric trait β, we consider two particular captures $b_{ref}, b_{search} \leftarrow \beta$ made at two different times. b_{ref} is the biometric data which is captured during the enrollment of a given user in the biometric system. This biometric data b_{ref} is stored to be used later as the data to be compared with. b_{search} corresponds to the "fresh" data which has to be matched against the references in the system in order to identify users.

There are many matching algorithms in the literature depending on the type of biometric data we are working with. The underlying performances of the recognition inside the biometric system is directly related to these algorithms.

Remark 1. *In a biometric system, we would like to minimize at the same time the False Accept Rate and the False Reject Rate which measures the probability that an impostor is wrongly identified and the probability that a genuine user is rejected [25, 30].*

In this chapter, we make the hypothesis that the biometric b are quantized and can be presented as binary vectors of length n, $b \in \{0, 1\}^n$ in such a way that:

Condition 1.

1. *Two different captures b, b' from the same user $\mathcal{U}$ are with high probability at a Hamming distance $d(b, b') \leq \lambda_{min}$.*
2. *Captures b_1, b_2 of different users $\mathcal{U}_1$, $\mathcal{U}_2$ are at a Hamming distance $d(b_1, b_2) > \lambda_{max}$.*

The Hamming distance $d(.,.)$ counts the coordinates in which two vectors differ. Condition 1 says that there good chances that $b, b' \leftarrow \beta$ are represented by binary vectors which are close, i.e. with many coordinates with the same value. On the other side, captures of different users should not have too many in common.

The origin of the hypothesis we made comes from iris recognition technology. Indeed [19] introduces the concept of iriscodes where the features of iris are represented as binary vectors of length $n = 2048$ which are compared with their Hamming distance. Following this, various attempts have been made for applying this kind of representations to fingerprints [26, 11] or more recently faces [17].

Remark 2. *[10] also makes use of this binary representation of biometric data for identification purpose. However, this work focuses on small database (typically, some hundred of users) and relies on the tamper resistance of some hardware component to achieve the confidentiality of the stored biometric data. This is not the case here.*

Condition 1 can be exploited to lead to efficient identification procedures as the one we are going to describe now. It should be noted that, still keeping iriscodes as an example, [23] introduces an efficient search algorithm for large database taking the same premises as those described here. The advantage of this way of proceeding is that it is possible to transpose it even with we have to deal with encrypted biometric data as we will show in Section 5.

The main difficulty we have to confront with is the variability of the captures of a biometric trait. In a nutshell, the idea to deal with the variability of the captures of the biometric data consists in comparing small parts of the binary vectors b_{ref} and b_{search} looking for an equality. Condition 1 says us that, on one hand, with a high probability the vectors we are comparing will have many small portions of their coordinates equal whenever they come from the same user. On the other hand, we should not be able to find such coincidences for two different users. By setting some threshold τ, this leads us directly to an algorithm:

1. Project b_{ref} (resp. b_{search}) onto l small portions of coordinates, to obtain m bits vector: $b_{ref_1}, \ldots, b_{ref_l}$ (resp. $b_{search_1}, \ldots, b_{search_l}$)
2. $c \leftarrow 0, j \leftarrow 1$
3. While $c < \tau$ and $j \leq l$
4. If $b_{ref_j} = b_{search_j}$ then $c \leftarrow c + 1$
5. $j \leftarrow j + 1$
6. End While
7. If $c = \tau$ then output 'matching'

Example 1. *Take iriscodes with $n = 2048$, $l = 128$ and $m = 10$. Let $\lambda_{min} = 0.25 \cdot 2048 = 512$, $\lambda_{max} = 0.35 \cdot 2048 = 716.8$ be the values of Condition 1. Note that $b_{ref_1}, \ldots, b_{ref_l}$ (resp. $b_{search_1}, \ldots, b_{search_l}$) are considered as independent vectors.*

We then have the probability p_1 to obtain the same small proportion for two iriscodes coming from the same user $p_1 = (1 - \frac{\lambda_{min}}{2048})^{10} \simeq 0.056$ and the probability to obtain the same value for two iriscodes coming from different users $p_2 = (1 - \frac{\lambda_{max}}{2048})^{10} \simeq 0.013$. With $\tau = 4$, then the probability not to output 'matching' for a legitimate user is

$$\sum_{i=0}^{\tau} \binom{l-1}{i} p_1^i (1 - p_1)^{l-i} \simeq 0.066$$

and the probability to output 'matching' for an impostor is

$$\sum_{i=\tau}^{l} \binom{l}{i} p_2^i (1 - p_2)^{l-i} \simeq 0.095.$$

Remark 3. *As explained above, a biometric identification system aims to recognize users by comparing fresh biometric templates to a set of enrolled ones. The system will output a small set of candidates (users) which are assumed to correspond to the fresh template. In theory, we would like to have either an empty set or a singleton. In practice, the accuracy of the system has a direct impact on the ratio between the size of the set and the size of the database.*

3. Security Model

In a biometric system, we have mainly three different components. The enrollment service is in charge of registering the users by capturing some biometric traits $b_i \leftarrow \beta_i$ for each user $\mathcal{U}_i$. The (central) database is storing the reference data which are computing from the enrolled biometric traits. An identity corresponding either to the real identity of the user or to a virtual identifier is also stored together with the reference data. The identification service is able to run an identification query towards the database with respect to the capture of a biometric trait. Depending on the context, the identification could lead to the output of the candidate identities or of the candidate templates.

Protection against external attacks is a classical issue. Such a system needs a secure implementation and infrastructure to mitigate external threat like eavesdropping, injection of malicious software, etc. For instance, establishing a secure channel ensuring confidentiality and authenticity during the communications between two distant components is important but already well-known. Consequently, we here focus on the internal threats, i.e. on what could be exploited or learned at the enrollment side, at the database side and at the identification service side.

The system should return an accurate result for the identification functionality. This leads to the two following requirements.

Requirement 1. *If the identification service is executing an identification query with respect to a template b_{search} coming from an enrolled user $\mathcal{U}_i$ (say a legitimate), then the identification service should output a list of candidates containing, except for a small probability, an index referring either to the enrolled template of $\mathcal{U}_i$ or to his identifier within the system.*

Requirement 2. *If the identification service is executing an identification query with respect to a template b_{search} coming from a non-enrolled user (say an impostor), then the identification service should output a list which is, except for a small probability, empty.*

For privacy issues, we want to ensure that the identification system is not misused and that function creep (i.e. use for another purpose) is prevented. This means in particular that a component should not be able to learn more information than what is really needed for a correct result. For instance, reference data in the central database should be handled in a protected form, an enrollment query or an identification query should not reveal private information to the database and the identification service should not retrieve information on the other data which are not related to the identification query. This is summarized in the requirements below.

The next requirement thwarts the situation when an adversary has recovered the content of the database and tries to gain some information.

Requirement 3 (Database Confidentiality). *The data which are stored in the central database should reveal information neither on the identities of the enrolled users nor on their biometric traits.*

The next requirement corresponds to the case where an adversary at the database side tries to detect whether a known biometric trait of a user is being registered, or to find a link between the identities or the traits.

Requirement 4 (Enrollment Privacy). *When executing enrollment queries, the database should not be able to distinguish them.*

The next requirement is the transposition of the previous one to the identification phase as we do not want such an adversary to be able to know which biometric data are looked for and which candidates are retrieved. In particular, this avoids to track a user through several identifications.

Requirement 5 (Identification Privacy). *When executing identification queries, the database should not be able to learn information about the content of the queries.*

The last requirement could apply at the identification service side. We do not want to let an adversary to learn more information from the result of an identification query than what is needed. This is not mandatory for all applications but seems to be a nice objective for non-criminal implementations of biometric identification system:

Requirement 6 (Identification Proportionality). *The information retrieved at the identification service side after executing an identification query for a template b_{search} should not give information on the rest of the database which is not related to b_{search}.*

4. Technical Ingredients

In this section, we introduce the techniques which are combined to achieve a secure identification. We will explain what is the goal of each technique and how they contribute to the construction of identification protocols.

4.1. Nearest Neighbor Problems

Let $\mathcal{P}$ be a set of points in a metric space (E, d_E). The Nearest Neighbor (NN) problem is for an input $x \in E$ to find a point $p_x \in \mathcal{P}$ such that

$$d_E(x, p_x) = d_E(x, \mathcal{P}) = \min_{p \in \mathcal{P}} d_E(x, p).$$

This topic has many applications in information retrieval. Efficient algorithms exist for searching a solution in low dimension space whereas this is not the case for large set of points in high dimensions [32]. The Approximate Nearest Neighbor (ANN) problem has been introduced to overcome this issue: For an input $x \in E$ and $\epsilon \geq 0$, find a point $p_x \in \mathcal{P}$ such that

$$d_E(x, p_x) \leq (1 + \epsilon) \min_{p \in \mathcal{P}} d_E(x, p).$$

Several algorithms have been proposed for the ANN problem [32]. The set of points $\mathcal{P}$ is preprocessed to create look-up tables enabling an efficient search through exact matches. Specific functions are selected and evaluated on the points; the tables are filled with the resulting values. Those functions should ensure that the evaluations on two close inputs would result most probably on the same output.

4.2. Locality-Sensitive Hashing

Locality-Sensitive Hashing (LSH) is a generic definition given in [33] for such functions. The basic idea is to map a point of a n dimension space into a value of a $m << n$ dimension space, following the principles of hash functions, with the requirement that it gives the same result – with a good probability – for near points. Note that the functions are thus different from cryptographic hash functions.

Definition 1 (Locality-Sensitive Hashing, [33]). *Let (E, d_E) be a metric space, U be a set, $r_1, r_2 \in \mathbb{R}$ with $r_1 < r_2$, $p_1, p_2 \in [0, 1]$ with $p_1 > p_2$. Let $\mathcal{H} = \{h_1, \ldots, h_l\}$ be a family of functions $h_i : E \to U$.*

The family $\mathcal{H}$ is (r_1, r_2, p_1, p_2)-LSH, if

$$\forall x, x' \in E \begin{cases} Pr_{h \in \mathcal{H}}[h(x) = h(x') \mid d_E(x, x') < r_1] > p_1 \\ Pr_{h \in \mathcal{H}}[h(x) = h(x') \mid d_E(x, x') > r_2] < p_2 \end{cases}$$

In practice, U is a space with a smaller dimensionality than E or a set of quite small size. For instance, if E is the Hamming space of dimension n, then U can be the Hamming space of a small dimension $m << n$. The principle of LSH is to reduce the differences between similar inputs up to a proximity threshold.

Good examples of LSH family, when E is a vector space, are to take random projections on small subspace (cf. [29, 28, 33, 2]). The family of functions used in [23] simply consist to restrictions of a 2048 bits vector to 10 of its coordinates (cf. Example 1).

With a LSH family $\mathcal{H} = \{h_1, \ldots, h_l\}$, a generic algorithm to solve the Approximate Nearest Neighbor (ANN) problem is as follows. For all $p \in \mathcal{P}$, preprocess the points to obtain look-up tables $T_1, \ldots, T_l$ which will be subsets of $\mathcal{P}$:

1. Let $T_1, \ldots, T_l$ be empty look-up tables with U as input domain.
2. For all $p \in \mathcal{P}$,
3.   For i from 1 to l do
4. $T_i[h_i(p)] \leftarrow \{p\} \cup T_i[h_i(p)]$
5.   End For
6. End For

To reduce the storage cost, we often store set of identifiers in the look-up tables where each identifier is related to a point of $\mathcal{P}$.

Finally to answer a query for a point $x \in E$, the algorithm computes $h_1(x), \ldots, h_l(x)$, recovers the corresponding subset of points $T_i[h_i(x)]$, $i \in \{1, \ldots, l\}$ and then analyses these subsets to output the point with the highest number of occurrences as the most probable nearest neighbor.

We retrieve above the principle exploited in [23] for iriscodes. The algorithm is similar and declares an iriscode b_{ref} of the central database as a potential candidate with respect to the input b_{search} as soon as b_{ref} (or its identifier) appears in the $T_i[h_i(b_{search})]$, $i \in \{1, \ldots, l\}$ more than τ times where τ is seen as a threshold. This algorithm does not offer any specific privacy and confidentiality properties as the data used to speed up the search for candidates are never encrypted. Nonetheless, the preprocessing step associated of such ANN search algorithm converts the approximate search step into an exact search issue.

4.3. Searchable Encryption

We introduce here the definition of cryptographic primitives that give a solution to deal with secure and exact searches. These primitives are related to the notions of Searchable Encryption [18] and Public-Key Encryption with Keyword Search [5]. The main goal is to store messages into an encrypted database while still enabling to search the messages related to some keywords. For instance this could correspond to a remote mailing service where the user wants to retrieve his messages which contain a given keyword, without letting the server learn information on the content of his mails.

Three entities are involved: a sender which wants to store data, a server that stores the data sent by the server, and a receiver which may retrieve part of the data detained by the server. Depending of the setting, the sender and the receiver could be distinct persons or the same person. In the latter case the construction usually relies on symmetric cryptography whereas asymmetric cryptography is used in the first case. The main functionalities are Send and Receive:

- Send$(m, \{\alpha_1, \ldots, \alpha_l\}, K)$ is a query run by the sender to store the message m in an encrypted format at the server side. $\{\alpha_1, \ldots, \alpha_l\}$ correspond to keywords associated to the message by the sender. The algorithm Send takes also as input a key K which is used to derive from m and the keywords the data that will actually be sent to the server.

- Receive(α, K') is a query run by the receiver, with inputs a keyword α and a key K' (possibly equal to K). The algorithm interrogates the server to retrieve the messages for which α is an associated keyword.

The two main security properties which are associated to searchable encryption schemes are the sender privacy and the receiver privacy. Namely, the sender privacy models the fact that the server should not learn information on what is stored at the server side (this always includes the confidentiality of the messages, the keywords and may encompasses the indistinguishability of the queries as well with asymmetric cryptography). The receiver privacy corresponds to the confidentiality of the receiver queries: the server should neither learn which data are retrieved and possibly nor distinguish the queries. Depending on the level of security, the scheme can prevent non-adaptive or adaptive attack strategies. The scheme [6] achieves the highest known level of security (indistinguishability, adaptive adversary) thanks to the use of Private Information Retrieval (PIR) [21] techniques. A PIR protocol enables to retrieve a specific block from a database without letting the database learn anything about the query and the answer (i.e. neither the index of the block nor the value of the block).

The general principle, in particular in [6, 18], to construct these primitives is to consider each sent keyword as a virtual address where the receiver can recover a link toward the associated messages. To do so without increasing too much the memory cost, several works [6, 22, 3] relies on Bloom filters.

4.4. Bloom Filters

Bloom filter is a notion used in membership checking applications to reduce the memory cost of the data storage.

Definition 2 (Bloom Filter, [4]). *Let $\mathcal{S}$ be a finite subset of a space E. For a family of ν (independent and random) hash functions $\mathcal{H}' = \{h'_1, \ldots, h'_\nu\}$, with each $h'_i : E \rightarrow \{1, \ldots, k\}$, a (ν, k)-*Bloom filter *is H, together with the array $(t_1, \ldots, t_k)$, defined as:*

$$t_\alpha = \begin{cases} 1 \textit{ if } \exists i \in \{1, \ldots, \nu\}, x \in \mathcal{S} \textit{ s.t. } h'_i(x) = \alpha \\ 0 \textit{ otherwise} \end{cases}$$

A Bloom filter enables to compress the representation of a dataset by introducing the possibility for some false positive to happen. Testing if an element x is in $\mathcal{S}$ is made by verifying that $t_(h'_i(x)) = 1$ for all i. So, there is no false negative and the probability for a false positive to happen is $\left(1 - (1 - \frac{\nu}{k})^{|\mathcal{S}|}\right)^{\nu}$.

An extension of this notion called *Bloom filters with storage* and defined in [6] enables to store an identifier of the related elements in each array:

1. $\forall i \in \{1, \ldots, k\}, t_i \leftarrow \emptyset$,
2. $\forall x \in \mathcal{S}$, $\forall j \in \{1, \ldots, \nu\}$, $t_{h'_j(x)} \leftarrow t_\alpha \cup \{Id(x)\}$ where $Id(x)$ is the identifier of x.

5. Secure Identification Protocols

5.1. Secure Identification Protocol: A General Construction

The tools described above enable to construct identification protocols that ensure confidentiality and privacy of the data and the queries. The overview of a generic construction is given below.

Let $b_{ref} \leftarrow \beta$ be a biometric template coming from a user to be enrolled. Let $\mathcal{H} = (h_1, \ldots, h_l)$ be a Locality-Sensitive Hashing family adapted to the biometric features. Let (Send, Receive) be the primitives of a searchable encryption scheme.

The **enrollment service** is constructed as the execution of the send query

$$\mathsf{Send}(Id(b_{ref}), \{1||h_1(b_{ref}), \ldots, l||h_l(b_{ref})\}, K)$$

where $Id(b_{ref})$ is an identifier associated to b_{ref}.

The **identification service** is designed as described in Table 1.

Optionally, the algorithm can also output the identifier of the retrieved reference data.

Table 1. Identification Service

1. Capture and extract a fresh biometric template $b_{search} \leftarrow \beta'$
2. Compute the fresh "keywords" $1||h_1(b_{search}), \ldots, l||h_l(b_{search})$
3. $\mathcal{C} \leftarrow \emptyset$, $c \leftarrow 0$, $j \leftarrow 1$
4. While $c < \tau$ and $j \leq l$
5. Execute the query $\mathsf{Receive}(i||h_i(b_{search}), K')$ to a receive a list L of identifiers
6. For each identifier x in L,
7. If $x \notin \mathcal{C}$ then initialize a counter $c_x \leftarrow 1$
8. Else $c_x \leftarrow c_x + 1$
9. $\mathcal{C} \leftarrow \mathcal{C} \cup \{L\}$
10. Assigned the maximum value of all assigned counters to c
11. $j \leftarrow j + 1$
12. End While
13. If $c = \tau$ then output 'success'

Concerning the accuracy requirements described in Section 3., we remark that the accuracy of the protocol depends mainly on the accuracy of the LSH family. For the security requirements, the properties of the searchable encryption scheme have clearly a direct impact. Nevertheless, the following examples illustrate that extra precautions are helpful, see below.

5.2. Survey of Existing Constructions

5.2.1. Asymmetric Identification Protocol

The first example is derived from [12] which introduced an error-tolerant searchable encryption scheme. An error-tolerant searchable encryption scheme enables a user to search messages with similar keywords. The construction follows the principle of the generic design above: use of LSH family combined with the searchable encryption scheme [6].

The principle is to combine LSH functions to the Bloom filter with storage so that the identification query consists in retrieving the identifiers attached to each keyword $i||h_i(b_{search})$. Due to Bloom filter structure, the keyword is evaluated by the ν Bloom filter hash functions, and then the intersection of ν retrieved tables is made to obtain one set of identifiers for each keyword. Then, with the l sets (one per keyword) of identifiers that have been retrieved, the algorithm just needs to count which identifier appears more than τ times.

All the identifiers are stored encrypted, ensuring the Database Confidentiality. The enrollment uses Private Information Storage (PIS) protocols. Whereas a PIR protocol enables to retrieve information of a database, a PIS protocol [31] enables to write information in a database while preventing the database from learning information on what is being stored (neither the value of the data, nor the index of the location where the data is being stored). Thus, thanks to PIS properties, the protocol achieves Enrollment Privacy. Moreover, the identification query is based on a PIR query to retrieve the information, which leads to Identification Privacy. Finally, PIS and PIR security properties hold against an adaptive adversary, so does the identification protocol.

In [12], the cost of an identification query is estimated. The main effort resides on the database side to answer to the PIR queries: indeed, the computation cost of the database for answering to a PIR query is linear in the size of the database. For one identification query, $l \times \nu$ PIR queries are executed (ν per keyword $i||h_i(b_{search})$).

5.2.2. Asymmetric Identification Protocol with Identification Proportionality

In [13], the scheme above have been extended to achieve Identification Proportionality. The problem with the previous scheme is that, among the results of the $l \times \nu$ PIR queries – which are intersected and analyzed to select the candidates – each result taken independently reveals additional information than what is necessary to construct the list of candidates.

For instance, it can give information on the similarity between users, even if they are not sufficiently close to be identified as a same user. To reach the Identification Proportionality requirement, the scheme from [12] is modified to enable the identification service to obtain information only when a identifier appears more times than a fixed threshold in the different queries. This is made possible thanks to specific secret sharing techniques where the basic goal is to split a secret s into n re-randomizable parts [34, 24].

5.2.3. Efficient Symmetric Identification Protocol

In [1], the generic principle of combining LSH with searchable encryption is applied to the symmetric searchable encryption from [18].

The benefit of symmetric searchable encryption compared to asymmetric schemes is the complexity (computation cost at the database side is much lower). However, a main disadvantage is the level of privacy. The existing symmetric searchable encryption schemes could suffer from statistical attacks on the database side: the database can distinguish the search patterns. This means first that the enrollment and identification services do not offer indistinguishability of the data but only confidentiality (the database can detect when a keyword – even if it remains unknown – was queried several times). Moreover, the database can also deduce links between the keywords and the messages from the relation between search patterns. Exploited in a biometric identification protocol, this leads to a privacy threat where the database would be able – after a kind of statistical analysis – to infer relations between users and their enrolled templates.

To prevent this, [1] modified [18] scheme by introducing an extra layer of encryption. This extra layer avoids the database guessing links between messages and keywords. At the end, this leads to a biometric identification protocol which achieves Database Confidentiality, Enrollment Privacy (when all enrollments are done in one time), and Identification Privacy against non-adaptive adversary and Identification Privacy with confidentiality only against an adaptive adversary.

A major difference with the two previous asymmetric identification protocols is on the computation cost. When the data are organized as look-up tables, a search could be made in constant time and only l searches (one per keyword $i||h_i(b_{search})$) are needed for an identification query.

6. Conclusion

We have described in this chapter the recent progress on the topic of biometric identification with encrypted data. Starting from the capture of a biometric trait, it has to be converted into a binary format which satisfies some properties (Condition 1). The situation of this conversion varies according to the kind of the considered biometric data. The representation already exists for iris, seems possible for faces and is the subject of on-going researches for the minutiae of fingerprints. Once done, these binary vectors have to be integrated into cryptographic primitives such as Public-key Encryptions with Keyword Search and Private Information Retrieval protocols.

There are today room for many improvements. On the biometric side, more progress are necessary to reach the accuracy of the traditional matching algorithms. Moreover, a reduction of the size of the binary biometric vector would be helpful. For cryptographic protocols, the primitives we suggest have to be improved a lot in terms of performances. They can not be easily used in practice today. Maybe, some trade-offs between the privacy properties and the speed of the implementation will have to be imagined.

We thus have to tackle down difficult problems in biometric and cryptography in order to obtain valuable results. However, we think that this is the price to pay to have solutions with good privacy properties for biometric identification; which constitutes in our mind the "real biometric experience".

References

[1] Michael Adjedj, Julien Bringer, Hervé Chabanne, and Bruno Kindarji. Biometric identification over encrypted data made feasible. In Atul Prakash and Indranil Gupta, editors, *ICISS*, volume 5905 of *Lecture Notes in Computer Science*, pages 86–100. Springer, 2009.

[2] Alexandr Andoni and Indyk Piotr. Near-optimal hashing algorithms for approximate nearest neighbor in high dimensions. *Commun. ACM*, **51**(1):117–122, 2008.

[3] John Bethencourt, Dawn X. Song, and Brent Waters. New constructions and practical applications for private stream searching (extended abstract). In *IEEE Symposium on Security and Privacy*, pages 132–139. IEEE Computer Society, 2006.

[4] Burton H. Bloom. Space/time trade-offs in hash coding with allowable errors. *Commun. ACM*, **13**(7):422–426, 1970.

[5] Dan Boneh, Giovanni Di Crescenzo, Rafail Ostrovsky, and Giuseppe Persiano. Public key encryption with keyword search. In Cachin and Camenisch [16], pages 506–522.

[6] Dan Boneh, Eyal Kushilevitz, Rafail Ostrovsky, and William. E. Skeith III. Public key encryption that allows PIR queries. In Alfred Menezes, editor, *CRYPTO*, volume 4622, pages 50–67. Springer, 2007.

[7] Julien Bringer and Hervé Chabanne. An authentication protocol with encrypted biometric data. In Serge Vaudenay, editor, *AFRICACRYPT*, volume 5023 of *Lecture Notes in Computer Science*, pages 109–124. Springer, 2008.

[8] Julien Bringer, Hervé Chabanne, Gérard D. Cohen, Bruno Kindarji, and Gilles Zémor. Theoretical and practical boundaries of binary secure sketches. *IEEE Transactions on Information Forensics and Security*, **3**(4):673–683, 2008.

[9] Julien Bringer, Hervé Chabanne, Malika Izabachène, David Pointcheval, Qiang Tang, and Sébastien Zimmer. An application of the goldwasser-micali cryptosystem to biometric authentication. In Josef Pieprzyk, Hossein Ghodosi, and Ed Dawson, editors, *ACISP*, volume 4586 of *Lecture Notes in Computer Science*, pages 96–106. Springer, 2007.

[10] Julien Bringer, Hervé Chabanne, Tom A. M. Kevenaar, and Bruno Kindarji. Extending match-on-card to local biometric identification. In Julian Fiérrez-Aguilar, Javier Ortega-Garcia, Anna Esposito, Andrzej Drygajlo, and Marcos Faúndez-Zanuy, editors, *COST 2101/2102 Conference*,

volume 5707 of *Lecture Notes in Computer Science*, pages 178–186. Springer, 2009.

[11] Julien Bringer, Hervé Chabanne, and Bruno Kindarji. The best of both worlds: Applying secure sketches to cancelable biometrics. *Sci. Comput. Program.*, **74**(1-2):43–51, 2008.

[12] Julien Bringer, Hervé Chabanne, and Bruno Kindarji. Error-Tolerant Searchable Encryption. In *International Conference on Communications*, June 2009.

[13] Julien Bringer, Hervé Chabanne, and Bruno Kindarji. Identification with encrypted biometric data. *Security Comm. Networks*, 2010. doi:10.1002/sec.206.

[14] Julien Bringer, Hervé Chabanne, David Pointcheval, and Qiang Tang. Extended private information retrieval and its application in biometrics authentications. In Feng Bao, San Ling, Tatsuaki Okamoto, Huaxiong Wang, and Chaoping Xing, editors, *CANS*, volume 4856 of *Lecture Notes in Computer Science*, pages 175–193. Springer, 2007.

[15] Julien Bringer, Hervé Chabanne, David Pointcheval, and Sébastien Zimmer. An application of the boneh and shacham group signature scheme to biometric authentication. In Kanta Matsuura and Eiichiro Fujisaki, editors, *IWSEC*, volume 5312 of *Lecture Notes in Computer Science*, pages 219–230. Springer, 2008.

[16] Christian Cachin and Jan Camenisch, editors. *Advances in Cryptology - EUROCRYPT 2004, International Conference on the Theory and Applications of Cryptographic Techniques, Interlaken, Switzerland, May 2-6, 2004, Proceedings*, volume 3027 of *Lecture Notes in Computer Science*. Springer, 2004.

[17] Chun Chen and Raymond N. J. Veldhuis. Binary biometric representation through pairwise polar quantization. In Massimo Tistarelli and Mark S. Nixon, editors, *ICB*, volume 5558 of *Lecture Notes in Computer Science*, pages 72–81. Springer, 2009.

[18] Reza Curtmola, Juan A. Garay, Seny Kamara, and Rafail Ostrovsky. Searchable Symmetric Encryption: Improved Definitions and Efficient

Constructions. In *CCS '06: Proceedings of the 13th ACM conference on Computer and communications security*, pages 79–88. ACM, 2006.

[19] John Daugman. High confidence visual recognition of persons by a test of statistical independence. *IEEE Trans. Pattern Anal. Mach. Intell.*, **15**(11):1148–1161, 1993.

[20] Yevgeniy Dodis, Leonid Reyzin, and Adam Smith. Fuzzy extractors: How to generate strong keys from biometrics and other noisy data. In Cachin and Camenisch [16], pages 523–540.

[21] William I. Gasarch. A survey on private information retrieval. http://www.cs.umd.edu/ gasarch/pir/pir.html.

[22] Eu-Jin Goh. Secure indexes. Cryptology ePrint Archive, Report 2003/216, 2003. `http://eprint.iacr.org/2003/216/`.

[23] Feng Hao, John Daugman, and Piotr Zielinski. A fast search algorithm for a large fuzzy database. *Information Forensics and Security, IEEE Transactions on*, 3(2):203–212, June 2008.

[24] Amir Herzberg, Stanislaw Jarecki, Hugo Krawczyk, and Moti Yung. Proactive secret sharing or: How to cope with perpetual leakage. In Don Coppersmith, editor, *CRYPTO*, volume 963 of *Lecture Notes in Computer Science*, pages 339–352. Springer, 1995.

[25] Anil K. Jain, Patrick Flynn, and Arun A. Ross. *Handbook of Biometrics*. Springer-Verlag New York, Inc., Secaucus, NJ, USA, 2007.

[26] Anil K. Jain, Salil Prabhakar, Lin Hong, and Sharath Pankanti. Fingercode: A filterbank for fingerprint representation and matching. In *CVPR*, pages 2187–. IEEE Computer Society, 1999.

[27] Ari Juels and Martin Wattenberg. A fuzzy commitment scheme. In *ACM Conference on Computer and Communications Security*, pages 28–36, 1999.

[28] Adam Kirsch and Michael Mitzenmacher. Distance-sensitive Bloom filters. In *Algorithm Engineering & Experiments*, Jan 2006.

[29] Eyal Kushilevitz, Rafail Ostrovsky, and Yuval Rabani. Efficient search for approximate nearest neighbor in high dimensional spaces. In *Symposium on the Theory Of Computing*, pages 614–623, 1998.

[30] Stan Z. Li and Anil K. Jain, editors. *Encyclopedia of Biometrics*. Springer US, 2009.

[31] Rafail Ostrovsky and Victor Shoup. Private information storage (extended abstract). In *STOC*, pages 294–303, 1997.

[32] Indyk Piotr. Nearest neighbors in high-dimensional spaces. In Jacob E. Goodman and Joseph O'Rourke, editors, *Handbook of Discrete and Computational Geometry, chapter 39*. CRC Press, 2004. 2rd edition.

[33] Indyk Piotr and Motwani Rajeev. Approximate nearest neighbors: Towards removing the curse of dimensionality. In *Symposium on the Theory Of Computing*, pages 604–613, 1998.

[34] Adi Shamir. How to share a secret. *Commun. ACM*, **22**(11):612–613, 1979.

[35] Koen Simoens, Pim Tuyls, and Bart Preneel. Privacy weaknesses in biometric sketches. In *IEEE Symposium on Security and Privacy*, pages 188–203. IEEE Computer Society, 2009.

[36] Qiang Tang, Julien Bringer, Hervé Chabanne, and David Pointcheval. A formal study of the privacy concerns in biometric-based remote authentication schemes. In Liqun Chen, Yi Mu, and Willy Susilo, editors, *ISPEC*, volume 4991 of *Lecture Notes in Computer Science*, pages 56–70. Springer, 2008.

In: Biometrics: Theory, Applications and Issues ISBN: 978-1-61728-765-7
Editor: Ellen R. Nichols, pp. 143-174

Chapter 7

THE EFFICIENCY OF SYSTEMATIC DESIGNS IN UNREPLICATED FIELD TRIALS

R.J. Martin[1], N. Chauhan[2], B.S.P. Chan[3] and J.A. Eccleston[4]

[1]Wirksworth, Derbys DE4 4EB, UK;
[2]Protherics plc, Runcorn, Cheshire, UK;
[3]ABN-Amro Bank, Hong Kong, China;
[4]School of Physical Sciences, University of Queensland, Brisbane, Australia.

Abstract

Systematic designs (where the allocation of varieties to plots is according to some fixed, non-random, pattern or scheme) are often used for the controls in large early generation variety trials (EGVTs). These are field trials which compare a large number of new varieties (or lines) with standard/control ones, and which are used to select some high yielding lines for further investigation. Usually the new lines are unreplicated at a site. The efficiency of different types of systematic unreplicated EGVT designs compared to optimal or efficient designs is investigated. Two design optimality criteria are used, and are compared for their association with the probability of selecting the highest yielding varieties. The data are assumed to be spatially dependent, and three models are considered for variety effects: both test and control effects fixed, controls effects fixed but variety effects random; and both test and control variety random.

1. Introduction

In early generation variety trials (EGVTs) there is a large number of new lines to be tested - usually greater than 200. These trials are usually what is called unreplicated, since at each site the new (or test) lines are unreplicated, usually due to the limited amount of seed available. Replicated standard varieties (checks or controls) are included for comparative purposes, and also to allow the detection of possible spatial trends. Several (often two or three) different sites may be used to investigate different environments. The purpose of the trials is to select top performers from the new lines, which are then to be tested further in later stages of the breeding programme. As well as the mean yield, other factors, such as quality, may also be of interest and affect selection. For further background information, see Patterson & Silvey (1980), Kempton (1984), and Kempton & Talbot (1988).

Recently there has been much interest in the spatial analysis of replicated and unreplicated field trials, where it is assumed that yields from field plots are spatially correlated, see, for example, Besag & Kempton (1986), Cullis *et al.* (1989), Martin (1990), Cullis & Gleeson (1991), Kempton *et al.* (1994), Grondona *et al.* (1996), Gilmour *et al.* (1997), Cullis *et al.* (1998), and Besag & Higdon (1999). However, little is known about the design of efficient unreplicated trials under spatial correlation. Some recent investigations are discussed in Martin, Chauhan et al. (2006), and Martin, Eccleston et al. (2006), denoted MCEC and MECC, respectively, henceforth.

Consider here a single site, with a two-dimensional layout of m = pq plots in p rows and q columns, and suppose that there are c control varieties and t test lines, giving v = c + t varieties in all. Let r_i denote the number of times that control variety i is replicated (i = 1, ..., c), denoted r under equi-replication.

Traditionally the experimental designs most frequently employed for EGVTs have the *control plots* (plots with a control variety) systematically distributed among the unreplicated test varieties, i.e. according to some non-random pattern or scheme. Some other designs are mainly systematic, but include a small amount of random allocation. Federer & Raghavarao (1975) give r×r augmented designs (ADs) where each control occurs once in each row and column. Lin & Poushinsky (1983) proposed modified ADs (MADs) for square or nearly square plots with $p=q$. Most of the controls are allocated systematically, and 17% to 13% of the total plots are control plots for $p=12$ to $p=36$. Lin & Poushinsky (1985) give MADs for long and narrow plots. Kempton (1984) recommended that the proportion of control plots 1 - t/m should be less than 1 in 5.

Kempton (1984) gave several examples of systematic designs, one of which has the control plots in diagonals three plots apart (see Figure 1(a)). Recently, designs with the control plots along separated diagonals have been used in EGVTs by New South Wales (NSW) Agriculture, Australia. The example in Federer (1998) with c=2 fills every third diagonal with controls, with mainly the same control in each diagonal (and t/m only 2/3). Another example in Kempton (1984), also in Besag & Kempton (1986), has the two controls filling every sixth column and alternating within the columns (see Figure 1(b)). This type of design is recommended for long and narrow plots, and has been used for winter wheat trials in the UK. Cullis *et al.*(1989) state that long narrow plots are generally used in cereal breeding programmes in Australia and that the most common layout has the controls filling columns at a given frequency, with additional control plots placed throughout the trial. They give an example of a NSW wheat trial with p=10, q=67, t=525, c=7 and $(r_1, ..., r_7)$ = (121, 3, 4, 5, 4, 4, 4). Control 1 filled every sixth column and the remaining control plots were allocated randomly. Kempton & Talbot (1988) give an example of a systematic design for roughly square plots with c=4, which has controls in every sixth column, but with the initial columns alternately being 1 or 4 (see Figure 1(c)).

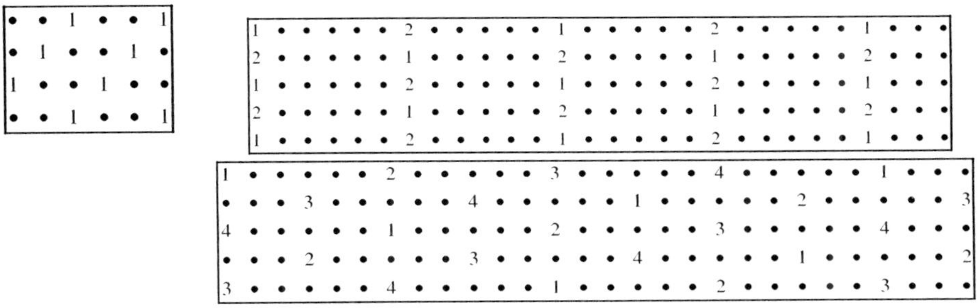

Figure 1. Examples of systematic layouts of controls (• represents the new varieties and numbers 1 to 4 the control varieties).

Some recent developments include Federer (2002), who has suggested using augmented lattice square designs, which were extended to α-α designs by Williams & John (2003); and Cullis et al. (2003) have generalised unreplicated EGVTs to p-rep ones, where a proportion p of the test lines are duplicated.

Recently, algorithmic methods have been developed to search for efficient designs when an underlying correlation structure is assumed. Martin & Eccleston (1997) developed a simulated annealing algorithm and demonstrated its applicability to various types of experiments using several examples, including unreplicated designs. See also Eccleston & Chan (1998).

Since the aim of an EGVT is to select the best new varieties for further testing, rather than estimate contrasts, Chauhan (2000, chapter 10) and MECC examine how well some simple design efficiency criteria correlate with selection probabilities. The robustness of unreplicated designs is also considered in MECC, and it is shown that the efficiencies of good designs under a specified variance structure are generally high under a different, but not too dissimilar, variance structure. MCEC provide some theoretical results on unreplicated designs using approximations to efficiency criteria values. The results given in MECC and MCEC assumed analyses with fixed variety effects, but there is increasing interest in modelling random variety effects. This has the often-desirable effect of shrinking the variety estimates (predictions) towards 0.

Federer & Raghavarao (1975), Patterson & Silvey (1980) and Cullis & Gleeson (1991) assumed fixed variety effects for plant breeding trials. Cullis *et al.* (1989) and Federer (1998) assumed fixed control effects and random new variety effects. Cullis *et al.* (1998), in their model for multi-environment EGVTs, assumed that both control and new varieties effects are random.

In this chapter, the efficiency and robustness of systematic designs are investigated under the assumptions that variety effects are fixed or random, under different correlation structures, and using two different efficiency criteria. The theory for comparing designs is given in Section 2. Some systematic designs for the two examples considered, and the methods of comparison, are given in Section 3. The efficiency and robustness of these designs is considered in Sections 4 and 5. Finally, the relevance of the two criteria to selection probabilities is discussed in Section 6.

2 Preliminaries

2.1. Models

We consider the representation

$$y = 1_m \mu + A\alpha + B\beta + Z_r \gamma + Z_c \delta + \xi + \varepsilon , \qquad (1)$$

where y is the $m \times 1$ vector of observations in lexicographic order; and α, β, γ, δ, ξ and ε are control variety effects, new variety effects, random row effects, random column effects, additional spatial effects, and random error, respectively, with A, B, $Z_r = I_p \otimes 1_q$ and $Z_c = 1_p \otimes I_q$ being the corresponding design matrices. We assume that $\theta = (\gamma', \delta', \xi', \varepsilon')'$ has zero mean, and the components are mutually uncorrelated, with γ, δ, and ε uncorrelated, and Λ is the correlation matrix of ξ, so that the variance matrix of θ is $\text{Var}(\theta) = \text{diag}\{\sigma_r^2 I_p, \sigma_c^2 I_q, \sigma_s^2 \Lambda, \sigma_e^2 I_m\}$. Let ψ denote (σ_r^2, σ_c^2, σ_s^2, σ_e^2).

It is commonly assumed for simplicity that the process for ξ is separable (Martin, 1990), so that $\Lambda = \Lambda_c \otimes \Lambda_r$, where Λ_c and Λ_r are the within-column and within-row correlation matrices, respectively. The spatial model used here is the separable AR1*AR1, which was shown by Grondona *et al.* (1996) to model the spatial component well in cereal yield trials. It is also recommended by Gilmour *et al.* (1997) and Cullis *et al.* (1998). Let λ_1 and λ_2 be the within-column and within-row parameters of the AR1*AR1 process, respectively, with $\eta = (\lambda_1, \lambda_2)$.

The different models which result from the control and test variety effects in (1) being assumed to be fixed or random are denoted Models 1, 2 and 3:

Model 1: This has α and β assumed to be fixed effects, and is the model considered in MECC and MCEC. It has $E(y) = \mu 1_m + A\alpha + B\beta$ and $\text{Var}(y) = V_1 = \sigma_r^2 (I_p \otimes J_q) + \sigma_c^2 (J_p \otimes I_q) + \sigma_s^2 \Lambda + \sigma_e^2 I_m$, where J_n is an $n \times n$ matrix of ones. Let X denote [A B]. Note that the rank of $[1_m\ X] = \text{rank}(X) = v$, and so μ can be omitted.

Model 2: This has β assumed to be a random effect, but α fixed, with $(\theta', \beta')'$ having zero mean and variance matrix $\text{diag}\{\text{Var}(\theta), \sigma_\beta^2 D_\beta\}$. Hence $E(y) = \mu 1_m + A\alpha$ and $\text{Var}(y) = V_2 = V_1 + \sigma_\beta^2 B D_\beta B' = V_1 + X'DX$, where $D = \text{diag}\{\sigma_\alpha^2 D_\alpha, \sigma_\beta^2 D_\beta\}$, and here $\sigma_\alpha^2 = 0$. The rank of $[1_m\ A]$ is c+1.

Model 3: This has both α and β assumed to be random effects, with $(\theta', \alpha', \beta')'$ having zero mean and variance matrix $\text{diag}\{\text{Var}(\theta), \sigma_\alpha^2 D_\alpha, \sigma_\beta^2 D_\beta\}$. Hence $E(y) = \mu 1_m$ and $\text{Var}(y) = V_3 = V_2 + \sigma_\alpha^2 A D_\alpha A' = V_1 + \sigma_\alpha^2 A D_\alpha A' + \sigma_\beta^2 B D_\beta B' = V_1 + X'DX$.

For Model 1, let σ_1^2 denote $\text{Var}(y_{i,j})$ and $\rho_{g,h}^{(1)} = \text{corr}(y_{i,j}, y_{i+g,j+h})$ be the lag (g, h) correlation, i.e. the correlation between two observations which are g rows and h columns apart. Then $\sigma_1^2 = \sigma_r^2 + \sigma_c^2 + \sigma_s^2 + \sigma_e^2$ and for g, $h > 0$

$\rho^{(1)}_{g,0} = \left(\sigma_c^2 + \sigma_s^2\lambda_1^g\right)/\sigma_1^2$, $\rho^{(1)}_{0,h} = \left(\sigma_r^2 + \sigma_s^2\lambda_2^h\right)/\sigma_1^2$, $\rho^{(1)}_{g,h} = \sigma_s^2\lambda_1^g\lambda_2^h/\sigma_1^2$. Models 2 and 3 are non-stationary. For Model 2, Var($y_{i,j}$) = σ_1^2 for control plots, and σ_1^2 + σ_β^2 for test plots, with Cov($y_{i,j}$, $y_{i+h,j+g}$) = $\rho^{(1)}_{g,h}$ σ_1^2. For Model 3, Var($y_{i,j}$) = σ_1^2 + σ_α^2 for control plots, and σ_1^2 + σ_β^2 for test plots, with Cov($y_{i,j}$, $y_{i+g,j+h}$) = $\rho^{(1)}_{g,h}$ σ_1^2 + σ_α^2 if both plots have the same control, and $\rho^{(1)}_{g,h}$ σ_1^2 otherwise.

2.2. Estimation snd Prediction of Variety Effects

For design purposes we assume that Var(y) is known, and that this matrix is to be used in generalized least-squares estimation and prediction. In practice, for data from a completed trial, a fitted variance structure will usually be used, but a design's efficiency should be similar if the assumed structure is a reasonable approximation.

Fixed effects can be estimated using the best linear unbiased estimator (BLUE), denoted $\hat{\alpha}$, while random effects can be predicted using the best linear unbiased predictor (BLUP), denoted $\tilde{\alpha}$. Let τ denote (α' β')'. Then in either case, the estimates or predictions can be found by, under Model k, solving an equation of the form $C_k\breve{\tau}$ = U'y for some U, and C_k of full rank ($\breve{\tau}$ contains the BLUE or the BLUP of α and β as appropriate). Then C_k is the 'information matrix' for τ, and, for Model 1 in the usual sense of estimating contrasts, $\text{var}(\breve{\tau}-\tau) = C_k^{-1}$. Let C_k^{-1} be partitioned into the control (1) and new (2) variety groups, that is, $C_k^{-1} = \begin{pmatrix} C_k^{11} & C_k^{12} \\ C_k^{12\prime} & C_k^{22} \end{pmatrix}$. Then, for example, C_k^{11} is $\text{Var}(\hat{\alpha})$ when k = 1, 2, and $\text{Var}(\tilde{\alpha}-\alpha)$ when k=3.

Consider the general full-rank mixed model E(y) = Fϕ + Hν + ε, with Var{(ν' ε')'} = diag{σ_ν^2D, $\sigma_\varepsilon^2 I_m$}, so that E(y) = Fϕ, Var(y) = V = σ_ν^2FDF' + $\sigma_\varepsilon^2 I_m$. Then $\hat{\phi}$ = $(F'V^{-1}F)^{-1}F'V^{-1}y$, and Henderson (1975) showed that the BLUP of ν is $\tilde{\nu}$ =

$\sigma_v^2 DH'V^{-1}(y - F\hat{\phi})$, and that these can be found from $\begin{pmatrix}\hat{\varphi}\\ \tilde{v}\end{pmatrix} = C^{-1}\begin{pmatrix}F'V^{-1}y\\ H'V^{-1}y\end{pmatrix}$,

where $C = \begin{bmatrix} F'V^{-1}F & F'V^{-1}H \\ H'V^{-1}F & H'V^{-1}H + (\sigma_\beta^2 D)^{-1}\end{bmatrix}$.

For Model 1, the estimator of τ can be taken as $\hat{\tau} = C_1^{-1}X'V_1^{-1}y$ where $C_1 = X'V_1^{-1}X$, with $\mathrm{Var}(\hat{\tau}) = C_1^{-1}$, as in MCEC.

For Model 2, adjusting for the constant μ gives $\hat{\alpha} = \left(A'V_2^{*}A\right)^{-1} A'V_2^{*}y$, and $\tilde{\beta} = \sigma_\beta^2 D_\beta B'V_2^{*}(y - A\hat{\alpha})$, where for a non-singular $n\times n$ matrix W, W^{*} denotes $W^{-1} - (1_n'W^{-1}1_n)^{-1}W^{-1}J_nW^{-1}$. These can be found using $\begin{pmatrix}\hat{\alpha}\\ \tilde{\beta}\end{pmatrix} = C_2^{-1}\begin{pmatrix}A'V_1^{*}y\\ B'V_1^{*}y\end{pmatrix}$, where $C_2 = \begin{bmatrix} A'V_1^{*}A & A'V_1^{*}B \\ B'V_1^{*}A & B'V_1^{*}B + (\sigma_\beta^2 D_\beta)^{-1}\end{bmatrix}$ is of full rank. Then $C_2^{11} = \left(A'V_2^{*}A\right)^{-1}$, $C_2^{12} = -\sigma_\beta^2 C_2^{11}A'V_2^{*}BD_\beta$, and $C_2^{22} = \sigma_\beta^2\left(D_\beta - \sigma_\beta^2 D_\beta B'V_2^{*}BD_\beta + \sigma_\beta^2 D_\beta B'V_2^{*}AC_2^{11}A'V_2^{*}BD_\beta\right)$. The Model 1 estimator $\hat{\tau}$ results as the limit of $\breve{\tau}$ as $\sigma_\beta^2 \to \infty$.

For Model 3, $\tilde{\tau} = C_3^{-1}X'V_1^{*}y$, where $C_3 = X'V_1^{*}X + D^{-1}$ is of full rank, and $C_3^{-1} = D - DX'V_3^{*}XD$. The Model 2 estimator/predictor $\breve{\tau}$ results as the limit of $\tilde{\tau}$ as $\sigma_\alpha^2 \to \infty$.

Cullis *et al.* (1989) state that for most EGVTs, the test lines are either genetically independent or there is insufficient knowledge of the pedigrees, so $D_\beta = I_t$ is assumed here. Cullis *et al.* (1998) assume that the random control effects have the same distribution as the random test treatment effects. Hence $D_\alpha = I_c$ is also assumed here. Also $\sigma_\alpha^2 = \sigma_\beta^2 = \sigma_\omega^2$ is assumed, where σ_ω^2 denotes the *genetic variance*. Thus, henceforth $V_2 = V_1 + \sigma_\omega^2 BB'$, and $V_3 = V_2 + \sigma_\omega^2 AA'$.

For variety trials the number of test treatments, *t*, is usually large, and hence calculating the $(c+t)\times(c+t)$ matrices C_k^{-1} (k = 1, 2, 3) makes algorithmic searches for good designs slow. For Model 1, MCEC show how to reduce the problem to the inversion of much smaller matrices of sizes $m-t$ and c. For Models 2 and 3, a search algorithm appears to need to invert the m×m matrices V_2 and V_3 for each design considered, since V_2 and V_3 depend on the design layout (i.e. on A and B). However, using a similar approach to that given in the Appendix of Cullis *et al.* (1998), only matrices of sizes $m-t$ and c need to be inverted. Number the control plots from 1 to m-t, and let U be the $m \times (m-t)$ control-plot incidence matrix, with $U_{i,j} = 1$ if plot i is the jth control plot, and 0 otherwise; and let $V_4 = V_1 + \sigma_\beta^2 I_m$. Then $V_2^{-1} = V_4^{-1} + V_4^{-1} U G^{-1} U' V_4^{-1}$, and $V_3^{-1} = V_2^{-1} - V_2^{-1} A F^{-1} A' V_2^{-1}$, where $G = (1/\sigma_\beta^2) I_{m-t} - U' V_1^{-1} U$ and $F = (1/\sigma_\alpha^2) I_c + A' V_2^{-1} A$. Then V_4 only needs to be inverted once, and G and F are of sizes m-t and c, respectively.

2.3. Design Efficiency Criteria

The four ways given by Federer & Raghavarao (1975) to assess the efficiency of unreplicated designs correspond to comparing the average variance of all pairwise comparisons

i. among control varieties (A_{ss}-value),
ii. among new varieties (A_{nn}-value),
iii. between control and new varieties (A_{ns}-value), and
iv. among all varieties (A-value).

MCEC note that the four criteria values are linearly related:

$$v(v-1)\times(\text{A-value}) = 2ct\times(A_{ns}\text{-value}) + t(t-1)\times(A_{nn}\text{-value}) + c(c-1)\times(A_{ss}\text{-value}),$$

and that for c small and t large, the A- and A_{nn}-values will be very similar. Simulations (MECC) confirm that the A_{ss}-values are not relevant for this situation. Hence, only the A_{nn}- and A_{ns}-values will be considered here. Ignoring $\mathrm{Var}(y_{i,j})$, these can be expressed as, respectively,

$$\{2/(t-1)\}\left\{\mathrm{tr}(C_k^{22}) - t^{-1} 1_t' C_k^{22} 1_t\right\},$$

and

$$(ct)^{-1}\left\{t\times tr(C_k^{11})+c\times tr(C_k^{22})-2\times 1_c{}'C_k^{12}1_t\right\}. \tag{2}$$

An efficient design with respect to a criterion is one which has a relatively low value - close to the best known value; while an inefficient design is one with a relatively high value. Note that there are no simple lower bounds available for these criteria.

Since high-yielding varieties are selected for further trials, a good design criterion will associate well with the probability of selection. Although the A_{nn}-criterion appears to be more relevant, MCEC show that for Model 1 its use can give less stable good designs as the model assumptions change. Simulations of the selection probabilities can be used to help see if, or when, one criterion is preferable. Let $\pi_{k,z}$ be the probability that the z new varieties with the largest estimated yields include the k best new varieties ($k \leq z \leq t$). MECC describe, for Model 1, an example with p=20, q=8, t=140, c=2, r=10, where $\pi_{7,35}$ is estimated over some different variance and η settings using a range of good, poor, and random designs. Results for some further simulations with different settings for this example, and for another 20×8 layout with c = 5, (r_1, ..., r_5) = (6, 6, 6, 6, 2) (as in Example 2 here) are given in Chauhan (2000, chapter 10). In all these examples using Model 1, both criteria values correlate well (negatively) with the selection probabilities, with the A_{nn} slightly better - the main differences being for some inefficient designs which have similar A_{nn}-values but different A_{ns}-values. Whether one criterion is preferable will be considered further in Section 6 for Model 1, and also for Models 2 and 3, using the chosen systematic designs.

3. The Systematic Designs Used and Their Efficiencies

Two examples are considered in this paper. Both are small compared to some designs used in practice, but their size allows them to be examined in more depth. Example 1, also used in Eccleston & Chan (1998) and MECC, has controls equi-replicated. It has p=10, q=5, t=40, c=2 and r=5, with controls on 20% of the plots. Example 2 is a larger trial with unequally replicated controls. It has p=20, q=8, t=134, c=5, and (r_1, ..., r_5) = (6, 6, 6, 6, 2), with 16.25% control plots. Controls 1 to 4 represent standard varieties with respect to yield, and control 5 represents a standard variety with respect to quality. This reflects current practice at NSW

Agriculture, where a standard variety to assess quality, with fewer replications than the standard varieties to assess yield, is included.

Most of the systematic designs used here can be categorised as *Diagonal Designs* (DDs), in which the control plots are allocated along non-adjacent diagonals (see Figure1(a)); *Column Designs* (DCs), in which the controls fill columns (as in Fig 1(b)) or *Row Designs* (DRs), in which the controls fill rows; or *Knight's Move Designs* (DKs), where the control plots are at least a knight's move apart. For DDs, DCs and DRs, like controls may appear together in strings or separated. There are also some *Edge Designs* (DEs), which have controls only in the edge rows and columns. Note that any design formed from another by rotation or reflection, or permutation of variety labels among varieties with the same replicate numbers, is equivalent in that it will have the same criteria values as the original design. In the non-spatial case when $\sigma_s^2 = 0$, this also holds under any row or any column permutation.

Long narrow plots are used in many EGVTs (Cullis *et al.*, 1989), so the (Model 1) correlation of the yield of adjacent plots is likely to be higher when they have a common longer side. It is assumed here that the rows are closer together than the columns, so that $\rho_{1,0}^{(1)}/\rho_{0,1}^{(1)} \geq 1$ is likely. The following settings of $\eta = (\lambda_1, \lambda_2)$ and $\psi = (\sigma_r^2, \sigma_c^2, \sigma_s^2, \sigma_e^2)$ are considered here:

$$\eta_1 = (0.5, 0.5), \eta_2 = (0.6, 0.2), \eta_3 = (0.9, 0.1);$$

and

$$\psi_1 = (0, 0, 1, 0), \psi_2 = (1, 6, 6, 2)/8, \psi_3 = (5, 5, 6, 4)/10, \psi_4 = (1, 1, 0, 1).$$

The *spatial settings* have $\sigma_s^2 > 0$: ψ_1 represents the *purely spatial setting*, ψ_2 has $\sigma_c^2 > \sigma_r^2 > 0$ to reflect the situation with long thin plots, and ψ_3 has $\sigma_c^2 = \sigma_r^2 > 0$. The η settings reflect equal spatial dependence in both directions (η_1), or increasingly stronger relative dependence down the columns (η_2 and η_3). The *non-spatial setting* ψ_4 has no spatial component and identical error, row and column variances. For Models 2 and 3, the genetic variance σ_ω^2 also needs to be specified. Without loss of generality, the sum of the error and spatial variance components, $\sigma_s^2 + \sigma_e^2$, is taken as 1, and, to reflect what is likely in practice, the ratio of σ_ω^2 to these, $\sigma_\omega^2/(\sigma_s^2 + \sigma_e^2)$, (i.e. σ_ω^2), is taken as 0.1, 0.5 and 1. For convenience, the settings of each of η_1 to η_3 with each of ψ_1 to ψ_3 are labelled cases 1 to 9 (cases 1 to 3 have ψ_1 to ψ_3 with η_1, cases 4 to 6 have them with η_2, and cases 7 to 9 have them with η_3), and the non-spatial ψ_4 is labelled case 10. The ratios of the Model 1 lag one column to row correlations, $\rho_{1,0}^{(1)} / \rho_{0,1}^{(1)}$, are given in Table 1 for cases 1 to 9. The ranges of possible corr($y_{i,j}$, $y_{i+1,j}$)/corr($y_{i,j}$, $y_{i,j+1}$) for Model 2 and Model 3 when σ_{ω}^2 = 1 are given in Table 2.

Table 1. Spatial case numbers (left), and (right) values of $\rho_{1,0}^{(1)} / \rho_{0,1}^{(1)}$ (to 2 d.p.)

	ψ_1	ψ_2	ψ_3		ψ_1	ψ_2	ψ_3
η_1	1	2	3		1	2.25	1
η_2	4	5	6		3	4.36	1.39
η_3	7	8	9		9	7.13	1.86

Table 2. Ranges of possible corr($y_{i,j}$, $y_{i+1,j}$)/corr($y_{i,j}$, $y_{i,j+1}$) for Model 2 (left), and Model 3 (right) when σ_{ω}^2 = 1 (to 2 d.p.)

	ψ_1	ψ_2	ψ_3		ψ_1	ψ_2	ψ_3
η_1	(0.5, 2)	(1.47, 3.45)	(0.67, 1.5)		(0.33, 3)	(0.75, 4.25)	(0.44, 2.25)
η_2	(1.5, 6)	(2.85, 6.69)	(0.92, 2.08)		(0.5, 8)	(1.06, 8)	(0.53, 3)
η_3	(4.5, 18)	(4.65, 10.93)	(1.24, 2.79)		(0.82, 19)	(1.19, 12.13)	(0.67, 3.64)

For Model 1, MCEC show that most DRs are likely to be A_{nn}-efficient if the lag 1 within-column correlation $\rho_{1,0}^{(1)}$ is dominant, and those with no like-control adjacencies are likely to be A_{ns}-efficient if $\rho_{1,0}^{(1)}/\rho_{0,1}^{(1)}$ is large. MCEC also show that designs with the controls well separated, such as DKs, are likely to be both A_{nn}- and A_{ns}-efficient if $\rho_{1,0}^{(1)}$ and $\rho_{0,1}^{(1)}$ are of similar size and dominant, and A_{ns}-efficient if $\rho_{1,0}^{(1)}$ is dominant but $\rho_{1,0}^{(1)}/\rho_{0,1}^{(1)}$ is not so large. Theory suggests (see MCEC) that DCs will be inefficient under Model 1 with $\rho_{1,0}^{(1)} / \rho_{0,1}^{(1)} \geq 1$.

For Example 1, 17 systematic designs are considered. Designs $D_1$1 to $D_1$7 and $D_1$14 to $D_1$17 are shown in Figure 2. $D_1$1 and $D_1$2 are DDs with diagonals maximally separated, and with like- and unlike-control diagonal adjacencies, respectively. Designs $D_1$3 to $D_1$5 are DKs. Designs $D_1$6 to $D_1$13 are DRs. Let DR-$_u$(i, j) denote a row design filling rows i and j, with u equal to r if the controls are repeated (1 in row i, 2 in row j), and a if alternating (in rows and columns). Then $D_1$6 to $D_1$13 are, respectively, $DR_r(1, 10)$, $DR_a(1, 10)$, $DR_r(2, 9)$, $DR_a(2, 9)$, $DR_r(3, 8)$, $DR_a(3, 8)$, $DR_a(4, 7)$ and $DR_a(5, 6)$. Similarly, if $DC_u(i)$ denotes a column design filling column i, with u equal to r or a as above, then $D_1$14 to $D_1$17 are $DC_r(3)$, $DC_a(3)$, $DC_r(5)$, $DC_a(5)$, respectively. Designs $D_1$6, $D_1$7, $D_1$16, $D_1$17 are also DEs. Note that some of these designs, e.g. $D_1$13 to $D_1$17, and the DR_rs under the A_{ns}-criterion, were not expected to perform too well under Model 1 (MCEC), and might not be considered for use in practice. They were included here to investigate how the possible systematic designs differ, and because the theory for Model 1 may not be a useful guide for Models 2 and 3. In the non-spatial case 10, the DDs and DKs $D_1$1 to $D_1$5 are all equivalent (one control in each row, and both in each column), as are the DR_rs ($D_1$6, $D_1$8, $D_1$10), the DR_as ($D_1$7, $D_1$9, $D_1$11 to $D_1$13), and the DCs ($D_1$14 to $D_1$17).

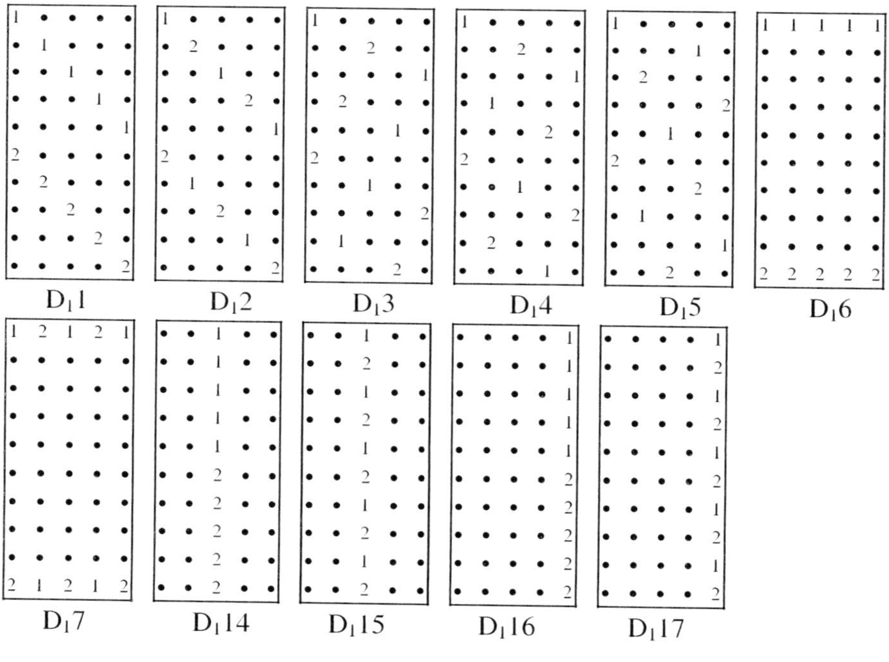

Figure 2. Example 1: Systematic designs $D_1$1 to $D_1$7 and $D_1$14 to $D_1$17 (• represents the new varieties and numbers 1 and 2 the control varieties).

For Example 2, the 21 designs in Figure 3 are considered. Designs $D_2$1 to $D_2$5 are DDs. $D_2$1 and $D_2$2 have no like-control diagonal adjacencies, whereas $D_2$3 and $D_2$4 have four like-control adjacencies for each of control varieties 1 to 4. For

$D_2$5, each of these controls is in a separate diagonal, so that each has 5 like-control adjacencies. Some variations on diagonal designs are considered here. $D_2$6 and $D_2$7 have all but two of the control plots in 12 diagonal pairs, with like and unlike pairs, respectively. $D_2$8 has the control plots mostly in 3 'V-shapes', with no like-control diagonal adjacencies (but 22 unlike ones). $D_2$9 to $D_2$14 are DKs, with the control plots at least lag (2, 1) apart for $D_2$9, $D_2$10, $D_2$13 & $D_2$14, and at least lag (1, 2) apart for $D_2$11 & $D_2$12. Designs $D_2$1, $D_2$3, $D_2$9, $D_2$11 and $D_2$13 have control 5 on opposite corner plots. For the other DDs and DKs, variety 5 is in the inner part of the array, with $D_2$5 having a diagonal self-adjacency. All these DDs and DKs have binary columns (i.e. no variety repeated) except $D_2$2, which has non-binary edge columns. Designs $D_2$15 to $D_2$18 are DEs, and $D_2$19 and $D_2$20 are similar, but have controls in the first two and last two rows. It is not possible to have perfect DRs in this example - the nearest would be 3 complete rows with two plots in another row, as in $D_2$21.

In Example 1, the 10 cases are considered for each of Models 1 to 3. Models 2 and 3 have three settings for σ^2_ω, giving 30 combinations in all. Since Example 2 is larger, the results from Example 1 (see section 4.2) were used to reduce the number of situations considered. So, in Example 2, only Models 1 and 3, and just η_1 and η_2, are considered. For Model 3 in Example 2, ψ_2 and σ^2_ω = 0.5 were not used, so Example 2 used 7 combinations for Model 1 (cases 1 to 6, and 10), and 10 for Model 3 (cases 1, 3, 4, 6, and 10).

Given the settings of ψ, η, and σ^2_ω, efficient designs under each of the two criteria can be found using search algorithms, and then compared over all settings. Here, the best design for each criterion was found by using the algorithm of Martin & Eccleston (1997) over some random starting designs for each case, each σ^2_ω, and each model; and then comparing those found. Example 1 used 100 initial designs, so that each best design (A_{nx}-best for x = n or s) should be very efficient for its settings, if not optimal. The algorithm takes considerably longer for Example 2 since the size of the array and the number of control plots are 3.2 and 2.6 times greater, respectively. Thus, fewer starting designs were used, and the best designs found are less likely to be optimal designs. Of the systematic designs considered, the best under the A_{nx}-criterion (x = n or s) is said to be *A_{nx}-s-best*; and others are said to be *A_{nx}-near-s-best* if they have A_{nx}-efficiency at least 99.5% of the efficiency of the A_{nx}-s-best design. Similarly, the *A_{nx}-s-worst* is the A_{nx}-worst systematic design considered.

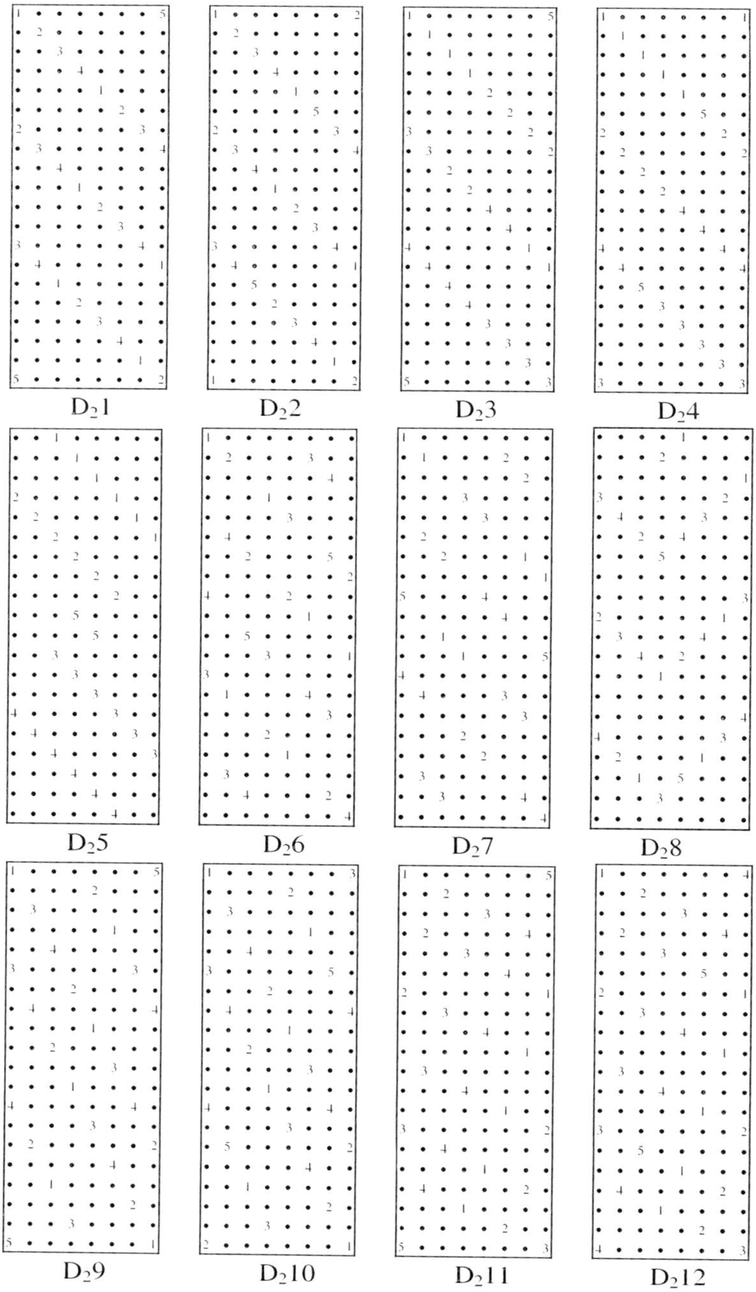

Figure 3. Continued on next page.

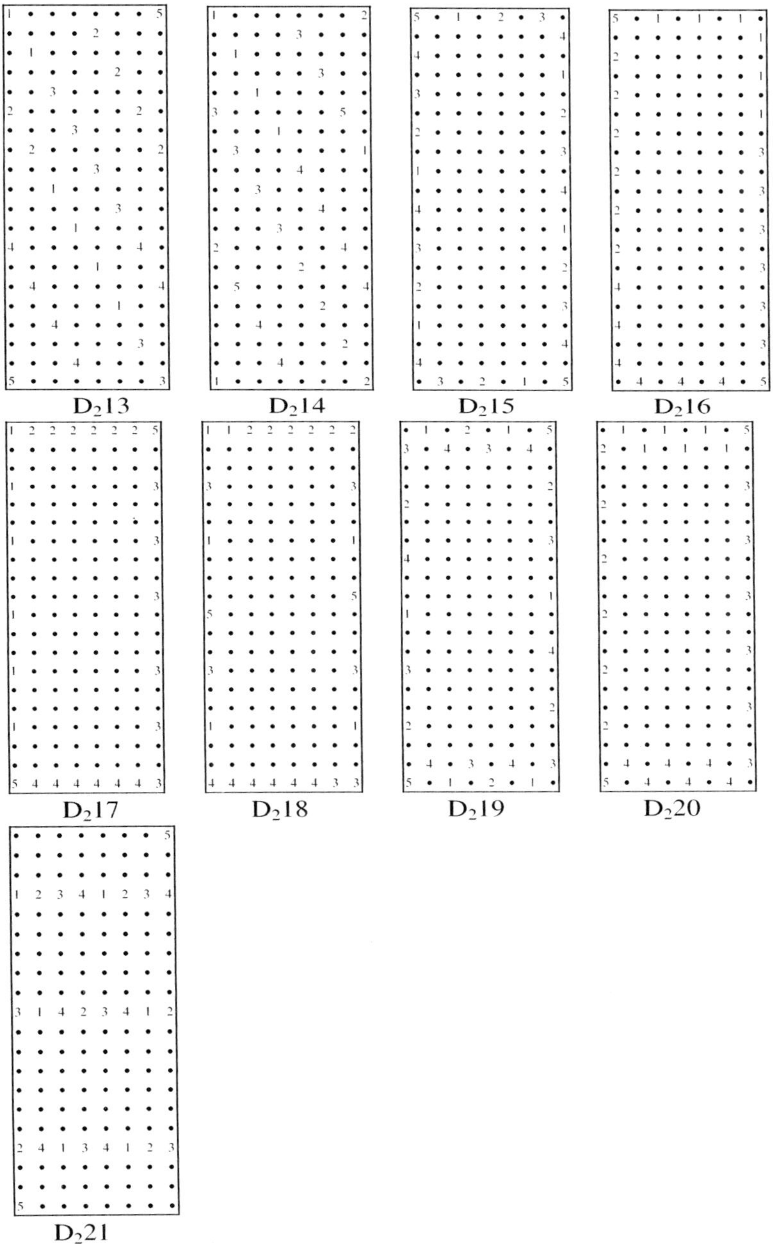

Figure 3. Example 2: Systematic designs $D_2 1$ to $D_1 21$ controls (• represents the new varieties and numbers 1 to 5 the control varieties).

To consider the robustness of the systematic designs in the spatial settings (cases 1 to 9), assume that there is no prior information on the variance components, other than that $\sigma_s^2 > 0$, and that η depends mainly on the plot dimensions. We discuss robustness for 6 possible kinds of prior information on plot shapes, depending on η, which we refer to as (a) near-square plots (η_1 - cases 1 to 3); (b) long thin plots (η_2 - cases 4 to 6); (c) very long thin plots (η_3 - cases 7 to 9); (d) not very long thin plots (η_1, η_2 - cases 1 to 6); (e) not square plots (η_2, η_3 - cases 4 to 9); (f) no prior information (η_1, η_2, η_3 - cases 1 to 9). For a particular prior assumption, a systematic design is called *m_1-robust* if the minimum efficiency over the cases considered (and over σ_s^2) is greater than m_1, and *(m_1, m_2)-robust* if the median efficiency is also greater than m_2.

4. Efficiency and Robustness of the Systematic Designs for Example 1

The results are given separately for Model 1 in Section 4.1, and Models 2 and 3 in Section 4.2. For each model, each case, and each σ_ω^2, the best and s-best designs were found under the two criteria (see Section 3). The best designs found are given in Appendix 1 ($BD_1 1$ to $BD_1 40$). The performance (efficiency and robustness) of the systematic designs $D_1 1$ to $D_1 17$ is evaluated here.

4.1. Model 1

4.1.1. Best Designs and Efficiency

Consider first the spatial cases 1 to 9 for Model 1. All the best designs found have binary columns except for the A_{nn}-best in case 1 ($BD_1 1$), which has control 1 twice in one row and three times in another, and similarly, in two different columns, for control 2. In the spatial cases 1 to 9, control plots in the best designs are at least lag 3 apart in columns (often at least lag 5 apart). Many best designs are symmetric (about a half-turn or a horizontal or vertical reflection). In cases 1 and 3, when $\rho_{1,0}^{(1)} = \rho_{0,1}^{(1)}$, the best designs have many of the control plots at least a knight's move apart. As expected from MCEC, when $\rho_{1,0}^{(1)} > \rho_{0,1}^{(1)}$, one of the DRs is A_{nn}-best ($D_1 10$ for cases 6 & 9; $D_1 11$ for cases 2, 4, 5 & 7; and $DR_r(3,9)$ for case 8), and the A_{ns}-best designs have many of the control plots at least a knight's move apart, although D_I11 is A_{ns}-best for case 7. Many of the other A_{ns}-best designs have at least one row adjacency between controls 1 and 2 (cases 2, 4, 5, 6, 8, 9). Also, when $\rho_{1,0}^{(1)} > \rho_{0,1}^{(1)}$, the A_{ns}-best designs do not have controls in the first or last rows.

Table 3 gives the s-best and nearly-s-best designs under each criterion, and the efficiencies of the s-best. Their efficiencies are all very high, and, as expected, D_I11 is s-best or near-s-best in cases 2 and 4 to 9 for the A_{nn}-criterion, and cases 4, 7, 8 for the A_{ns}-criterion. In the other cases, D_I4 is s-best or near-s-best. The two criteria give the same s-best or near-s-best designs for ψ_1, but there are some differences in the other cases, with the A_{nn}-criterion often preferring DRs (cases 2, 5, 6, 9).

Table 3. s-best and near-s-best designs (D_I omitted) for Example 1, Model 1, and (underneath) the efficiency of the s-best designs

	ψ_1		ψ_2		ψ_3	
	A_{nn}	A_{ns}	A_{nn}	A_{ns}	A_{nn}	A_{ns}
η_1	4, 5	4, 5	11	2, 4	2-5	2-4
	0.9944	0.9976	1	0.9903	0.9999	0.9988
η_2	11	11	10, 11	2, 4-5	10, 11	2-5
	1	0.9924	1	0.9843	1	0.9969
η_3	11	11	8-11	2, 11	8-11	2-5
	1	1	0.9985	0.9761	1	0.9822

The A_{nn}-s-worst design among D_I1 to D_I13 is D_I13 for cases 1 to 8, with respective efficiencies 0.842, 0.924, 0.934, 0.884, 0.894, 0.950, 0.654, 0.885. In case 9, it is D_I1, with efficiency 0.932. If D_I13 is excluded, the s-worst efficiencies are 0.854 (D_I6), 0.952 (D_I6), 0.948 (D_I6), 0.900 (D_I6), 0.928 (D_I6), 0.965 (D_I7), 0.782 (D_I6), 0.892 (D_I1), for cases 1 to 8, respectively. Apart from cases 1 and 3, the DCs D_I14 to D_I17 all have low A_{nn}-efficiencies - around 0.3 for case 7.

Among D_I1 to D_I13, the A_{ns}-s-worst design is always D_I6, with efficiencies 0.646, 0.752, 0.670, 0.794, 0.819, 0.722, 0.687, 0.830, 0.716 for cases 1 to 9, respectively. If the DR_rs (D_I6, D_I8, D_I10) and D_I13 are excluded, then D_I7 is s-worst, with usually much higher A_{ns}-efficiencies: 0.859, 0.957, 0.888, 0.851, 0.885, 0.826, 0.729, 0.905, 0.830, respectively. The remaining DR_as (D_I7, D_I9, D_I11, D_I12) only do well in cases 4, 5, 7, 8, which have a high $\rho_{1,0}^{(1)} / \rho_{0,1}^{(1)}$. Apart from case 1, when D_I15 performs similarly to D_I9 and D_I12, the DCs D_I14 to D_I17 all have very low A_{ns}-efficiencies - below 0.2 for case 7.

Of the two DDs, $D_1 2$ is always better than $D_1 1$ under both criteria, but the differences are fairly small (largest for case 1). The three DKs ($D_1 3$ to $D_1 5$) have very similar efficiencies, as do the four DCs ($D_1 14$ to $D_1 17$). The DRs show some interesting differences. For a pair of DRs ($D_1(2i)$ $D_1(2i+1)$), i = 3,4,5, with control plots in the same rows, the DR_r ($D_1(2i)$) has a lower A_{ns}-efficiency than the DR_a ($D_1(2i+1)$). The relative efficiency varies from 0.833 (case 1) to 0.942 (case 7). However, these designs have similar A_{nn}-values, with $D_1(2i+1)$ slightly more efficient in cases 6, 8, 9. As expected, $D_1 13$ is the A_{nn}-worst DR, and the A_{ns}-worst DR_a. Overall, $D_1 11$ is the best DR, but $D_1 10$ is also A_{nn}-good (and sometimes slightly A_{nn}-better). Also, $D_1 8$ and $D_1 9$ can be quite A_{nn}-good, and $D_1 9$ is A_{ns}-good throughout.

For the non-spatial case 10, the A_{nn}- and A_{ns}-best designs have controls evenly spread over columns, but have some rows with two unlike controls (as suggested in MCEC). The DDs and DKs are s-best under both criteria (A_{ns}-efficiency 0.9986, A_{nn}-efficiency 0.995). The DCs are equally A_{nn}-s-worst. The DRs have a high A_{nn}-efficiency of 0.975, but are less A_{ns}-good - the DR_rs have efficiency 0.691, and the DR_as have efficiency 0.807.

4.1.2. Robustness of the Systematic Designs

Now consider the robustness of the systematic designs $D_1 1$ to $D_1 17$. Over the 9 spatial cases, $D_1 2$ to $D_1 5$ have quite high efficiencies under both criteria (minimum 0.882). The DRs $D_1 8$ to $D_1 12$ have quite high A_{nn}-efficiency in all cases (minimum 0.853), but even the DR_as ($D_1 9$, $D_1 11$, $D_1 12$) have fairly low A_{ns}-efficiency when $\rho^{(1)}_{1,0} = \rho^{(1)}_{0,1}$ (0.813 to 0.857). All of $D_1 1$ to $D_1 12$ are reasonably A_{nn}-robust, with efficiency greater than 0.78. The (m_1, m_2)-robust designs (cases 1 to 9) for (m_1, m_2) = (0.85, 0.95), (0.9, 0.98) and (0.9, 0.99) are given in Table 4. The most robust A_{nn}-designs for near-square plots (shape (a)) are $D_1 4$ and $D_1 5$, and otherwise it is $D_1 11$. Designs are less robust under the A_{ns}-criterion, but $D_1 2$, $D_1 4$, $D_1 5$ are generally quite robust.

Table 4. (m_1, m_2)-robust designs (D_1 omitted) for cases 1 to 9 for Example 1, Model 1

Shape	Criterion	(0.85, 0.95)	(0.9, 0.98)	(0.9, 0.99)
(a)	A_{nn}	2-5, 8-12	3-5	4-5
	A_{ns}	2-5	2-5	4-5
(b)	A_{nn}	1-5, 8-12	5, 8-11	10-11
	A_{ns}	1-5, 9, 11-12	2, 4-5	none

Table 4. Continued

Shape	Criterion	(0.85, 0.95)	(0.9, 0.98)	(0.9, 0.99)
(c)	A_{nn}	8-12	8-11	8-11
	A_{ns}	1-5, 11	none	none
(d)	A_{nn}	2-5, 8-12	3-5, 9, 11	11
	A_{ns}	2-5	2-5	4-5
(e)	A_{nn}	8-12	8-11	10-11
	A_{ns}	1-5, 11	none	none
(f)	A_{nn}	2-5, 8-12	9, 11	11
	A_{ns}	2-5	2	none

4.2. Models 2 and 3

4.2.1. Best Designs and Efficiency for Model 2

Consider first the spatial cases 1 to 9 for Model 2. The A_{nn}-best design is one of the systematic designs $D_1 2$, $D_1 6$, $D_1 7$, $D_1 9$, $D_1 14$, $D_1 16$ in 25 of the 27 combinations. It is $D_1 6$ for 19 of them: all cases when $\sigma^2_\omega = 0.1$; cases 2, 4, 5, 6, 8, 9, when $\sigma^2_\omega = 0.5$; and cases 5, 6, 8, 9, when $\sigma^2_\omega = 1$. It is $D_1 9$ for case 7 when $\sigma^2_\omega = 0.5$ or 1; and when $\sigma^2_\omega = 1$ it is $D_1 7$ for cases 2, 4; and $D_1 2$ for case 3. For case 3 when $\sigma^2_\omega = 0.5$, it is $D_1 16$. For case 1 when $\sigma^2_\omega = 0.5$, it is a DE, and when $\sigma^2_\omega = 1$ the controls are fairly well separated with two (unlike) diagonal adjacencies.

The only systematic A_{ns}-best design for Model 2 is $D_1 11$ for case 7 when $\sigma^2_\omega = 1$. Otherwise, the A_{ns}-best designs have many of the control plots at least a knight's move apart. However, for ψ_1 with $\sigma^2_\omega = 0.1$, the A_{ns}-best designs have many like-control diagonal adjacencies, with like-controls clustered together (but no row or column adjacencies) away from the top and bottom edge plots. Many of the A_{ns}-best designs have at least one row or column adjacency between controls 1 and 2. The same best design ($BD_1 21$) arises for case 8 when $\sigma^2_\omega = 0.1$, and cases 2 and 5 when $\sigma^2_\omega = 0.5$; and similarly ($BD_1 22$) for case 9 when $\sigma^2_\omega = 0.1$, and cases 5 and 6 when $\sigma^2_\omega = 1$.

Table 5 gives the s-best and nearly-s-best designs under each criterion. When σ_ω^2 = 0.1, all of $D_1$1 to $D_1$17 have extremely high A_{nn}-efficiencies (at least 0.988) except for case 7, when they are all still very high (at least 0.946). For σ_ω^2 = 0.5, 1, the A_{nn}-efficiencies for $D_1$1 to $D_1$12 are all high - greater than 0.95, and 0.92, respectively. Apart from in case 7, $D_1$13 to $D_1$17 are only slightly worse. $D_1$6 is A_{nn}-s-best or near-s-best in every combination except for cases 1 and 7 when σ_ω^2 = 1, with $D_1$7 having very similar values. One of the DRs is A_{nn}-s-best in all but 4 combinations: $D_1$2 is s-best in case 1 when σ_ω^2 = 0.5, and cases 1 and 3 when σ_ω^2 = 1; and $D_1$16 is s-best in case 3 when σ_ω^2 = 0.5. The A_{nn}-s-worst design when σ_ω^2 = 0.5, 1 is one of $D_1$13 to $D_1$17, although these are only really bad for case 7.

Table 5. s-best and near-s-best designs (D_1 omitted) for Example 1, Model 2

σ_ω^2	Criterion	Case								
		1	**2**	**3**	**4**	**5**	**6**	**7**	**8**	**9**
0.1	A_{nn}	6, 7, 16, 17	All	All	6, 7	2-10, 14-17	All	6, 7	6, 7	All
	A_{ns}	1	1	1	1	1	1	11, 12	1	1
0.5	A_{nn}	2-7	2, 6, 7	2-11, 16, 17	6, 7	6, 7	6-8, 16	6-9	6, 7	6, 7
	A_{ns}	1, 3	4, 5	1, 4, 5	1-5	1	1	11	1	1
1	A_{nn}	2-5	2-9	1-8, 14, 16, 17	2, 4-9	6, 7	2-10, 16	8, 9	6-8	2-8
	A_{ns}	2-5	2, 4, 5	4, 5	2-5	1, 2, 4, 5	1	11	1, 2, 4, 5	1

There are major differences in A_{ns}-efficiency between $D_1$1 to $D_1$5 and the DRs and DCs. All of $D_1$1 to $D_1$5 have at least reasonably high A_{ns}-efficiencies in all combinations - the lowest is 0.853, although none is ever best. Apart from case 7, the A_{ns}-s-best design is $D_1$1 in 16 combinations (all 8 when σ_ω^2 = 0.1), $D_1$5 in 7 (5 when σ_ω^2 = 1), and $D_1$3 once. Their A_{ns}-efficiencies range from 0.962 to 0.995. The A_{ns}-s-best DRs are $D_1$11 and $D_1$12, with $D_1$9 also close. However, apart from case 7, when $D_1$11 (σ_ω^2 = 0.5, 1) or $D_1$12 (σ_ω^2 = 0.1) is A_{ns}-s-best overall, and case 4 with σ_ω^2 = 0.5, 1, the best A_{ns}-efficiency of $D_1$11 and $D_1$12 is 0.90, and it can be as low as 0.40 (case 3, σ_ω^2 = 0.1). As for Model 1, $D_1 6$ is the A_{ns}-s-worst among $D_1 1$ to $D_1 13$, with some very low efficiencies (ranging from 0.244 to 0.811), but the DCs are often even worse (e.g. efficiencies less than 0.1 for cases 7 and 8 when σ_ω^2 = 0.1). $D_1 8$ and $D_1 10$ are also usually quite bad (except for case 7).

As for Model 1, $D_1 2$ is always slightly A_{nn}-better than $D_1 1$, and there is little difference between the DKs. However, $D_1 1$ is A_{ns}-better than $D_1 2$ in 17 combinations. The difference in their A_{ns}-efficiencies is usually fairly small, but for case 1 with σ_ω^2 = 0.1, the A_{ns}-efficiencies are 0.962 ($D_1 1$) and 0.853 ($D_1 2$).

For the non-spatial case 10, any DR with each control in each column (such as $D_1 6$ to $D_1 13$) is A_{nn}-best for σ_ω^2 = 0.1, and any DC (such as $D_1 14$ to $D_1 17$) is A_{nn}-best for σ_ω^2 = 0.5, 1. Any design with one control in each row, and with each control in each column (such as $D_1 1$ to $D_1 5$) is A_{ns}-best for σ_ω^2 = 0.1, 0.5, 1. The other systematic designs all have extremely high A_{nn}-efficiencies - at least 0.9974. However, they have very low A_{ns}-efficiencies: 0.234, 0.357, 0.433 for the DR_rs when σ_ω^2 = 0.1, 0.5, 1, respectively; 0.357, 0.504, 0.583 for the DR_as; and 0.212, 0.328, 0.401 for the DCs.

4.2.2. Robustness of the Systematic Designs for Model 2

The (m_1, m_2)-robust designs (here also over σ_ω^2 = 0.1, 0.5, 1) are given in Table 6 for (m_1, m_2) = (0.9, 0.99), (0.95, 0.995) and (0.95, 0.9975) (A_{nn}); or (0.9, 0.95) and (0.95, 0.975) (A_{ns}). Clearly, $D_1 6$ and $D_1 7$ are the most A_{nn}-robust over the prior assumptions considered, and $D_1 1$ is usually the most A_{ns}-robust.

Table 6. (m_1, m_2)-robust designs (D_1 omitted) for cases 1 to 9 for Example 1, Model 2

	A_{nn}			A_{ns}	
Shape	**(0.9, 0.99)**	**(0.95, 0.995)**	**(0.95, 0.9975)**	**(0.9, 0.95)**	**(0.95, 0.975)**
(a)	1-12, 14, 16, 17	2-8	6, 7	1	1
(b)	2-12, 14-17	6, 7	6, 7	1-5	1
(c)	6-9	6, 7	6, 7	1-5	4, 5
(d)	1-12, 14, 16, 17	2, 4-8	6, 7	1	1
(e)	2-10	6, 7	6, 7	1-5	none
(f)	1-12	6-8	6, 7	1	none

4.2.3. Results For Model 3

With only two controls, results for Model 3 are, as might be expected, very similar to those for Model 2. For 47 times out of 60, the best designs are the same as for Model 2- see Appendix 1. The A_{nn}-best designs differ 6 times from Model 2, and the A_{ns}-best 7 times, but the efficiency differences between those best under the two models are very small. The s-best and nearly-s-best, given in Table 7, are very similar to those in Table 5.

Table 7. s-best and near-s-best designs (D_1 omitted) for Example 1, Model 3

σ_ω^2	Criterion	Case								
		1	2	3	4	5	6	7	8	9
0.1	A_{nn}	6, 7, 16, 17	All	All	6, 7	1-10, 14-17	All	6, 7	6, 7	All
	A_{ns}	1	1	1, 4, 5	1	1	1	11, 12	1	1
0.5	A_{nn}	2-7	2, 6, 7	1-9, 16, 17	6, 7	6, 7	6-8, 16	6-9	6, 7	6, 7
	A_{ns}	1-5	2-5	1, 4, 5	1-5	1, 4, 5	1	11	1	1
1	A_{nn}	2-5	2-9	1-7, 14, 16, 17	6-8	6, 7	1-10, 16	8, 9	6-8	2-8
	A_{ns}	2-5	2, 4, 5	4, 5	2-5	1-5	1, 4, 5	11	1, 2, 4, 5	1

The A_{nn}-efficiencies of the systematic designs are much the same as for Model 2, but the A_{ns}-efficiencies are not as low (at least 0.65). The differences between the efficiencies for Models 2 and 3 are given in Appendix 1. The robust designs, given in Table 8, are similar to those in Table 6.

Table 8. (m_1, m_2)-robust designs (D_1 omitted) for cases 1 to 9 for Example 1, Model 3

	A_{nn}			A_{ns}	
Shape	**(0.9, 0.99)**	**(0.95, 0.995)**	**(0.95, 0.9975)**	**(0.9, 0.95)**	**(0.95, 0.975)**
(a)	1-12, 14-17	2-7	6, 7	1	1
(b)	2-11, 14, 16, 17	6, 7	6, 7	1-5	1
(c)	6-9	6, 7	6, 7	1-5	4, 5
(d)	1-12, 14, 16, 17	2, 4-8	6, 7	1	1
(e)	2-10	6, 7, 8	6, 7	1-5	none
(f)	1-11	6-8	6, 7	1	none

5. Efficiency and Robustness of Systematic Designs for Example 2

5.1. Model 1

The seven best designs (cases 1 to 6, and 10) for Model 1 were found under each criterion - see Appendix 2 ($BD_2 1$ to $BD_2 14$). These have similar properties to the best designs in Example 1. The designs have binary columns except under ψ_1 (both criteria), and the A_{nn}-best under ψ_4. However, the A_{nn}-best designs for cases 2, 4, and 5 have some row clustering of the control plots (using 11, 12, and 5 rows, respectively) - that for case 5 is close to a DR. For these 3 cases, the lag 1 within-column correlation is dominant, with $\rho_{1,0}^{(1)} / \rho_{0,1}^{(1)} \geq 2.25$ - see Table 1. For the other cases (1, 3, 6), with $\rho_{1,0}^{(1)} / \rho_{0,1}^{(1)} \leq 1.39$, the control plots tend to be at least a knight's move apart, as in Example 1. The A_{nn}-best design for case 3 has 10 diagonal adjacencies (7 separated) of unlike controls, similarly to $D_2 6$, and one of like controls.

In the spatial cases 1 to 6, the A_{nn}-best designs have no column adjacencies between controls, but all the A_{ns}-best designs contain the arrangement $\begin{array}{|cc|} c_1 & 5 \\ 5 & c_2 \end{array}$, where $c_1 \neq c_2$ are two of the control varieties 1 to 4. All also have at least one other control c_3 next to one of these plots - usually next to control variety 5. This arrangement seems counterintuitive, but the theory in MCEC shows that it reduces the sum of the diagonal elements of C_1^{-1} corresponding to the controls. The A_{ns}-best designs also contain other row or column adjacencies between unlike controls.

The s-best and nearly-s-best designs, and the efficiency of the s-best, are given in Table 9. All of $D_2 1$ to $D_2 14$ are reasonably A_{nn}-efficient with A_{nn}-efficiency greater than 0.90, and one or both of $D_2 8$ (V shapes), $D_2 9$ (a DK) is A_{nn}-s-best or near-s-best in 5 cases. The DEs $D_2 15$ and $D_2 16$ are usually the A_{nn}-s-worst, but not exceptionally bad (the lowest efficiency is 0.79 for $D_2 16$ in case 5). The A_{ns}-efficiencies of $D_2 1$ to $D_2 14$ are at least 0.82, but those of $D_2 15$ to $D_2 21$ can be rather worse (the lowest is 0.69 for $D_2 17$ in case 5). The A_{nn}- and A_{ns}-efficiencies of $D_2 1$ to $D_2 14$ are highly correlated (greater than 0.93) in all 7 cases (including case 10).

Table 9. s-best and nearly-s-best designs (D_2 omitted), and (underneath) the efficiency of the s-best for Example 2, Model 1

	ψ_1		ψ_2		ψ_3	
	A_{nn}	A_{ns}	A_{nn}	A_{ns}	A_{nn}	A_{ns}
η_1	9, 10	6, 10	8, 9	8	1, 6, 9, 13	1, 13
	0.983	0.945	0.987	0.962	0.992	0.968
η_2	8	8	21	8	1, 6, 8, 9, 13	1, 13
	0.987	0.959	0.997	0.968	0.992	0.972

For the non-spatial case 10, the best designs have the control plots spread out fairly evenly over the rows and columns, although the A_{ns}-best has none in three rows. The s-best is $D_2 13$, with efficiency 0.986 under both criteria. The near-s-best are $D_2 3$ under both criteria, and also $D_2 1$ and $D_2 6$ under the A_{nn}-criterion. The A_{nn}-s-worst is $D_2 21$ and the A_{ns}-s-worst is $D_2 16$, with efficiencies 0.897 and 0.726, respectively.

For each pair of designs ($D_2 1$, $D_2 2$), ($D_2 3$, $D_2 4$), ($D_2 9$, $D_2 10$) and ($D_2 13$, $D_2 14$), the control plots used are identical, but the first has control 5 in two opposite corner plots, and the second has it inside the array. Over almost all the cases considered for these pairs of designs the first design in each pair is more efficient, suggesting that, generally, having control 5 in the corner plots gives a more efficient design. For each of the pairs ($D_2 1$, $D_2 3$), ($D_2 2$, $D_2 4$) and ($D_2 6$, $D_2 7$), the designs differ in that the first has unlike-control diagonal adjacencies and the second has like-control ones. A comparison of these designs suggests that, for the spatial cases, like-control diagonal adjacencies should be avoided. This is supported by $D_2 5$, which has the most like-control diagonal adjacencies, being the s-worst design among $D_2 1$ to $D_2 14$ in 4 cases (both criteria).

Since the number of cases considered here is smaller than for Example 1, only m_l-robustness is considered. The most robust designs overall (see Table 10) are $D_2 9$ (A_{nn}), and $D_2 6$, $D_2 10$ (A_{ns}).

Table 10. m_l-robust designs (D_2 omitted) for cases 1 to 6 for Example 2, Model 1

	A_{nn}		A_{ns}
Shape	**$m_l = 0.95$**	**$m_l = 0.975$**	**$m_l = 0.94$**
(a)	6-14	9, 10	6, 10
(b)	1-4, 6-14	1, 8, 9	1, 2, 6, 8-10
(d)	6-14	9	6, 10

5.2. Model 3

For Model 3, the 10 best designs (cases 1, 3, 4, 6, 10 with σ_{ω}^{2} = 0.1, 1) were found under each criterion - see Appendix 2 ($BD_2 15$ to $BD_2 31$). The limited searches carried out, and the lack of theory on good designs for this model, mean that specific features of these best designs may not be fully consistent with the features of optimal designs. Consider firstly the spatial cases. For σ_{ω}^{2} = 0.1, the A_{nn}-best designs are DEs. For cases 3 and 6, the best design has all the control plots on the edge columns (1 and 8) with one column full, and with like-controls quite clustered together (13 like column adjacencies, 10 unlike ones). For cases 1 and 4, like-control plots are clustered in the edge rows (1 and 20), with 12 row like-adjacencies, and the remaining controls are spread out in the edge columns. When σ_{ω}^{2} = 1, the A_{nn}-best designs have most of the control plots in the outer two or three rows and columns, with most control plots at least a knight's move apart for η_1 (cases 1 & 3), and many unlike-control diagonal pairs for η_2 (13 for case 4, and 12 for case 6).

However, the A_{ns}-best designs have few controls on the edges. For ψ_3 (cases 3 & 6), the A_{ns}-best designs are binary in rows and columns. For ψ_1 (cases 1 & 4) and σ_{ω}^{2} = 0.1, the controls form clusters or strings with many like diagonal adjacencies (overall, 22 and 27, respectively). For all cases with σ_{ω}^{2} = 1, and for case 1 with σ_{ω}^{2} = 0.1, the A_{ns}-best designs contain the same 2×2 arrangement for control 5 with two others as for Model 1, and, especially for ψ_1, have other row and column unlike-control adjacencies. For ψ_1 and σ_{ω}^{2} = 1, the A_{ns}-best designs have a large cluster of adjacent controls (10 for case 1, 13 for case 4).

The s-best and nearly-s-best designs are given in Table 11. All of $D_2 1$ to $D_2 21$ are highly A_{nn}-efficient, with A_{nn}-efficiency greater than 0.982 (0.993 for σ_{ω}^{2} = 0.1). As for Model 1, the designs with high numbers of diagonal like-control adjacencies ($D_2 3$, $D_2 4$, $D_2 5$, $D_2 7$) are worse than those with none, as are those with control 5 away from the corners rather than in them, but the differences are very small. The A_{ns}-efficiencies among $D_2 1$ to $D_2 14$ are at least 0.933, with those for $D_2 15$ to $D_2 21$ being at least 0.832 - at least 0.890 if $D_2 17$ and $D_2 18$ are excluded. Designs with many diagonal like-adjacencies are worse than those with unlike ones when σ_{ω}^{2} = 1, but better when σ_{ω}^{2} = 0.1. In most cases, having control 5 away from the corners is better. Again, the differences are very small. Comparing D_213 and D_214 with D_29 and D_210, respectively, there is a slight advantage under the A_{ns}-criterion in having the controls at least (1, 2) apart rather than at least (2, 1) apart.

Table 11. s-Best and near-s-best designs (D_2 omitted) for Example 2, Model 3

σ^2_ω	Criterion	η_1		η_2	
		ψ_1	ψ_3	ψ_1	ψ_3
0.1	A_{nn}	1-4, 6, 8-21	All	1-4, 6, 8-21	All
	A_{ns}	5	5, 8	5	3, 5
1	A_{nn}	1, 2, 6, 8-15, 19	1-15, 19-21	1, 2, 6, 8-14, 17-19, 21	1-15, 18, 19
	A_{ns}	2, 6, 8, 10, 12, 14	6, 8, 14	2, 6, 8, 10, 12, 14	5, 6, 8

As might be expected from the properties of the best designs, the efficiencies under the two criteria do not correlate well - highly negative for σ^2_ω = 0.1 (between -0.94 and -0.67), and at most 0.42 for σ^2_ω = 1.

For the non-spatial case 10 with σ^2_ω = 0.1, the A_{nn}-best design has the controls clustered in 5 rows. The A_{nn}-best design for σ^2_ω = 1, and the A_{ns}-best designs, have controls fairly evenly spread out, and are binary in both rows and columns. D_213 is the s-best under both criteria (efficiencies 1 or almost 1), but all efficiencies among D_21 to D_214 are at least 0.998 (A_{nn}) and 0.981 (A_{ns}) - higher when σ^2_ω = 0.1, and the criteria values correlate very highly (0.89 and 0.97 at σ^2_ω = 0.1, 1, respectively). Some of D_215 to D_221 are rather A_{ns}-worse (efficiency down to 0.854).

All of D_21 to D_221 are reasonably A_{nn}-robust. Table 12 gives the m_1-robust designs for m_1 = 0.95 and 0.96 (A_{ns}) or 0.965 (A_{nn}), with designs D_219 most A_{nn}-robust, and D_26 and D_28 the most A_{ns}-robust.

Table 12. m_1-robust designs (D_2 omitted) for cases 1, 3, 4, 6 for Example 2, Model 3

	A_{nn}		A_{ns}	
Shape	**m_1 = 0.95**	**m_1 = 0.965**	**m_1 = 0.95**	**m_1 = 0.96**
(a)	1, 2, 6, 8-15, 19	11, 15, 19	2, 6, 8, 10, 12, 14	6, 8
(b)	1, 2, 6, 8-14, 18, 19	19	1-14	2, 4-8, 10, 14
(d)	1, 2, 6, 8-14, 19	19	2, 6, 8, 10, 12, 14	6, 8

6. Relationships Between Selection Probabilities and Criteria Values

The results in Sections 4 and 5 show that there can be some important differences between efficient designs under the two criteria. Simulation studies (as in MECC) estimated the selection probability $\pi_{k,z}$ (see Section 2.3) from 10,000 sets of simulated yields for the systematic designs over the cases considered. For Example 1, k=2, z=10; and for Example 2, k=7, z=45. The standard error of each $\hat{\pi}_{k,z}$ is roughly 0.005. Note that the analyses used the postulated covariance structure.

Consider first Example 1. The possible values of $\pi_{2,10}$ depend on the case. For example, for Model 1 the maximum of $\hat{\pi}_{2,10}$ over D_I1 to D_I17 is 0.59, 0.48, 0.42, 0.59, 0.49, 0.41, 0.85, 0.61, 0.47, 0.33 for cases 1 to 10, respectively; and the range is 0.057, 0.080, 0.025, 0.087, 0.097, 0.038, 0.302, 0.204, 0.074, 0.013, respectively. For case 1, the maximum of $\hat{\pi}_{2,10}$ and its range over D_I1 to D_I17 for Model 2, σ_ω^2 = 0.1, 0.5, 1, and Model 3, σ_ω^2 = 0.1, 0.5, 1, are (0.33, 0.021), (0.67, 0.020), (0.82, 0.020), (0.32, 0.022), (0.67, 0.023), (0.82, 0.016), respectively. The maximum range for Models 2 and 3 over σ_ω^2 = 0.1, 0.5, 1 is 0.058, 0.045, 0.034, 0.020, 0.032, 0.023, respectively - all for case 7 except the first (case 6). The estimates are fairly similar in Models 2 and 3, apart from case 7 which has much higher $\hat{\pi}_{2,10}$ for Model 2, e.g. maxima of 0.90 and 0.66 when σ_ω^2 = 0.5.

For Model 1, Table 13 gives those designs with the highest $\hat{\pi}_{2,10}$, or at most 0.015 less, plus the correlations, over D_I1 to D_I13, of $\hat{\pi}_{2,10}$ with the A_{ns}- and A_{nn}-efficiencies. The correlation is larger for the A_{nn}-efficiency except for case 4, and four of the A_{ns}-correlations are negative. Overall, the designs with lowest $\hat{\pi}_{2,10}$ are D_I13 and the DCs, and those with the highest are D_I9 to D_I11. In most cases, the range in Models 2 and 3 is not much greater than 0.015, so that most of D_I1 to D_I17 have fairly similar selection probabilities, with no consistent patterns.

Table 13. Designs (D_1 omitted) with $\hat{\pi}_{2,10}$ within 0.015 of the highest $\hat{\pi}_{2,10}$, plus (underneath, in brackets) the correlations with the A_{nn}- and A_{ns}-efficiencies, respectively, for Example 1, Model 1

	ψ_1	ψ_2	ψ_3
η_1	3-5	2-6, 8-12	1-5, 8, 9, 11, 15
	(0.91, 0.88)	(0.86, -0.02)	(0.67, 0.54)
η_2	2, 3, 10-12	3, 4, 7-12	1-4, 6-12
	(0.91, 0.97)	(0.39, 0.30)	(0.38, -0.50)
η_3	9-11	6, 8-12	6, 7, 9-13
	(0.97, 0.90)	(0.75, -0.21)	(0.18, -0.27)

The $\hat{\pi}_{2,10}$ provide no evidence that DR_rs are worse than the corresponding DR_as, nor that the DDs and DKs are better in general than the DRs $D_1 6$ to $D_1 12$. They do show that the DCs are worse for Model 1, especially in cases 7 and 8. Overall, the results are more in accordance with the A_{nn}-values than the A_{ns}-values.

Now consider Example 2. For Model 1, there was little difference in $\hat{\pi}_{7,45}$ over $D_2 1$ to $D_2 14$. In 4 of the 7 cases, all were within 0.015 of the maximum, and the maximum difference over the other 3 cases was 0.026. The DEs, $D_2 15$ to $D_2 20$, usually had worse $\hat{\pi}_{7,45}$, but $D_2 21$ was within 0.015 for 4 cases (and had the highest value for case 5). For Model 3 with $\sigma_\omega^2 = 0.1$, the maximum $\hat{\pi}_{7,45}$ was very low (at most 0.075), and all the designs had $\hat{\pi}_{7,45}$ within 0.015 of the maximum. Most designs had close $\hat{\pi}_{7,45}$ when $\sigma_\omega^2 = 1$ (the maximum range over all 5 cases was 0.022). Again, the results give some support to using the A_{nn}-values rather than the A_{ns}-values.

Conclusion

The results for Model 1 were mainly as expected from the theoretical results in MCEC, with the best designs found, and the efficiencies of the systematic designs which were considered, dependent on the variance components ψ and the AR1*AR1 parameters η. However, for Models 2 and 3, as well as the settings of ψ and η, the properties of the best designs found are also dependent on the genetic variance σ_ω^2. The results for Models 2 and 3 in Example 1 are very similar, but, unlike for Model 1, it is difficult to ascertain theoretically properties of efficient designs for given ψ, η and σ_{ω}^{2}.

For many combinations of ψ and η, efficient designs under either criterion have the control plots fairly well separated. If the dependence is much stronger in one direction, then a Row or Column design may be better (especially under the A_{nn}-criterion). The results suggested that the A_{nn}-values may be better associated with selection probabilities than the A_{ns}-values. Although Edge Designs are A_{nn}-efficient for some combinations of ψ, η and σ_{ω}^{2}, practitioners may be reluctant to use such designs. Instead, Knight's Move designs, which have the control plots spread out, may be preferable, since they generally have both high A_{nn}- and A_{ns}-efficiencies.

When the range of the efficiencies of the systematic designs considered is very small, as for the A_{nn}-efficiencies in Example 2 under Model 3, and the range over randomly-selected designs is also small, comparisons of these designs may not be useful. The design could then be chosen on other grounds.

In summary, the results given here suggest that in many situations, a systematic design with the control plots as well separated as possible should work well. If there is high confidence that the necessary conditions hold, a Row or Column design could be considered, but in general, such designs should usually be avoided. It would also seem sensible to compare the criteria values for some possible proposed systematic designs with the values resulting from an algorithm which obtains efficient designs.

Since there is not a perfect relationship between the efficiency criteria used here and the selection probabilities, it could also be sensible in practice to compare the estimated selection probabilities of some possible designs.

In practice, the parameters of the variance structure would usually be estimated from the data. To estimate these parameters well, and if necessary, to assess the adequacy of the variance assumptions, it would usually be preferable to have some control plots close together and some more spread out. To take account of this, a systematic arrangement which has the control plots spread out over the array (such as a Knight's Move design), with some additional randomly allocated control plots or adjacent ones, may be useful.

Although it has been assumed here that the variance structure is known, and design robustness examined by looking at design efficiency over different variance structures, an alternative approach, used by Cullis et al. (2006), is to compare designs using simulated data and estimating all the parameters. This approach can be more realistic, but is dependent on the specific models used for the simulations, and on the specific methods used for estimation. The large amount of computation required also limits the number of designs that can be compared.

It was assumed here that it is reasonable to assume the test lines are genetically independent. In other cases, it may be possible to use information on the genetic relationships in the design (Bueno Filho & Gilmour, 2003).

A recent variation on unreplicated designs for comparing test lines has been proposed by Cullis et al. (2006). Their p-rep designs have a proportion p of the test lines duplicated, while the remainder are unreplicated. They show that these designs are more efficient than the systematic grid plot designs.

In practice, EGVTs on a given set of new lines are usually carried out at several different sites, perhaps selected to cover different soil types and local climates. It then seems desirable to design all the experiments together. Also, interest is usually not confined to the crop yield, with the crop quality also being assessed in a laboratory – a two-phase experiment. It can then be desirable that the overall design aims to ensure efficiency of the quality data as well as the yield.

The systematic and best designs are shown, and tables of criteria values are given in full, in Appendices 1 and 2, which are available from RJM.

Acknowledgment

We are grateful to B. Cullis of NSW Agriculture for many helpful discussions, and to the Australian Research Council for financial assistance. RJM is grateful for support provided by the Departments of Statistics at the University of Queensland, and for the facilities at the University of Otago. NC and BSPC are grateful to the EPSRC, UK and APA, Australia, respectively, for PhD studentships. We are also grateful for some detailed anonymous comments given on an earlier version.

References

Besag, J. and Kempton, R. (1986). Statistical analysis of field experiments using neighbouring plots. *Biometrics* **42**, 231-251.

Besag, J. and Higdon, D. (1999). Bayesian inference for agricultural field experiments (with discussion). *J. Roy. Statist. Soc. B* **61**, 691-746.

Bueno Filho, J. S. de S. and Gilmour, S. G. (2003). Planning incomplete block experiments when treatments are genetically related. *Biometrics* **59**, 375-381.

Chauhan, N. (2000). *Efficient and optimal designs for correlated observations*. Ph.D. thesis. University of Sheffield.

Cullis, B. R. and Gleeson, A. C. (1991). Spatial analysis of field experiments - an extension to two dimensions. *Biometrics* **47**, 1449-1460.

Cullis, B., Gogel, B., Verbyla, A. and Thompson, R. (1998). Spatial analysis of multi-environment early generation variety trials. *Biometrics* **54**, 1-18.

Cullis, B. R., Lill, W. J., Fisher, J. A., Read, B. J. and Gleeson, A. C. (1989). A new procedure for the analysis of early generation variety trials. *Appl. Statist.* **38**, 361-375.

Cullis, B. R., Smith, A. B. and Coombes, N. E. (2006). On the design of early generation variety trials with correlated data . *J. Agric. Biol. Env. Sci.* **11**, 381-393.

Eccleston, J. and Chan, B. (1998). *Design algorithms for correlated data*. In 'COMPSTAT 1998, Proceedings in Computational Statistics', Ed. Payne, R. and Green, P., pp. 41-52. Heidelberg: Physica-Verlag.

Federer, W. T. (1998). Recovery of interblock, intergradient, and intervariety information in incomplete block and lattice rectangle designed experiments. *Biometrics*, **54**, 471-481.

Federer, W. T. (2002). Construction and analysis of an augmented lattice square design. *Biometrical J.* **44**, 251-257.

Federer, W. T. and Raghavarao, D. (1975). On augmented designs. *Biometrics* **31**, 29-35.

Gilmour, A. R., Cullis, B. R. and Verbyla, A. P. (1997). Accounting for natural and extraneous variation in the analysis of field experiments. *J. Agric. Biol. Env. Statist.*, **2**, 269-293.

Grondona, M. O., Crossa, J., Fox, P. N. and Pfeiffer, W. H. (1996). Analysis of variety yield trials using two-dimensional separable ARIMA processes. *Biometrics* **52**, 763-770.

Henderson, C. R. (1975). Best linear unbiased estimation and prediction under a selection model. *Biometrics* **31**, 423-447.

Kempton, R. A. (1984). The design and analysis of unreplicated field trials. *Vortr. Pflanzenzüchtg.* **7**, 219-242.

Kempton, R. A., Seraphin, J. C. and Sword, A. M. (1994). Statistical analysis of two-dimensional variation in variety yield trials. *Journal of Agricultural Science* **122**, 335-342.

Kempton, R. A. and Talbot, M. (1988). The development of new crop varieties. *J. R. Statist. Soc. A* **151**, 327-341.

Lin, C. S. and Poushinsky, G. (1983). A modified augmented design for an early stage of plant selection involving a large number of test lines without replication. *Biometrics* **39**, 553-561.

Lin, C. S. and Poushinsky, G. (1985). A modified augmented design (type 2) for rectangular plots. *Canadian Journal of Plant Science* **65**, 743-749.

Martin, R. J. (1990). The use of time series models and methods in the analysis of agricultural field trials. *Commun. Statist. -Theor. Meth.* **19**, 55-81.

Martin, R. J., Chauhan, N., Eccleston, J. A. and Chan, B. S. P. (2006). Efficient experimental designs when most treatments are unreplicated. *Lin. Alg. Appl.* **417**, 163-182.

Martin, R. J. and Eccleston, J. A. (1997) *Construction of optimal and near-optimal designs for dependent observations using simulated annealing.* Research Report 479/97, Dept. Probab. Statist., Univ. Sheffield, UK. Also, Research Report #84, Dept. Maths., Univ. Queensland, Australia.

Martin, R. J., Eccleston, J. A., Chauhan, N. and Chan, B. S. P. (2006). Some results on the design of field experiments for comparing unreplicated treatments. *J. Agric. Biol. Env. Sci.* **11**, 394-410.

Patterson, H. D. and Silvey, V. (1980). Statutory and recommended list trials of crop varieties in the United Kingdom (with discussion). *J. Roy. Statist. Soc. A* **143**, 219-252.

Williams, E. R. and John, J. A. (2003). A note on the design of unreplicated trials. *Biometrical J.* **45**, 751-757.

INDEX

A

B

C

H

I

L

M

N

O

P

Q

R

S

T

U

V

W